# WHO BURNED AUSTRALIA?

'IT CAME like a ball of flame and went straight up the mountain. And it finished the town off.'

The town was tiny Mount Macedon in Victoria.

At 10 p.m. on February 16th, 1983 – Ash Wednesday – it was a living community – thirty minutes later a smoking ruin.

Mount Macedon's experience was repeated all over Australia as a four-year ecological disaster culminated in the worst bushfires in recorded history, fires one expert compared to a post-atomic fire storm.

Seventy-one people died. Property damage is measured in hundreds of millions of dollars.

Deaths and damage were the result of fires that in many cases were deliberately lit, were expected and predicted, could have been prevented and are certain to occur again.

Who burned Australia? The incredible answer is part of a worldwide mystery involving snowstorms in California and the feeding habits of the kangaroo, Mexican volcanic eruptions and some of the strangest plant life in the world.

In the best-selling *The Fire Came By*, written with T. R. Atkins, John Baxter investigated the riddle surrounding the Great Siberian Explosion of 1908. Now he turns his attention to a mystery no less extraordinary involving his own native continent. *Who Burned Australia?* is Baxter's twenty-second book. Among his other publications are six novels, including *The Hermes Fall* and *The Black Yacht*.

# WHO BURNED AUSTRALIA?

## John Baxter

NEW ENGLISH LIBRARY

A New English Library Original Publication, 1984

First NEL Paperback Edition 1984

NEL Books are published by
New English Library,
Mill Road, Dunton Green,
Sevenoaks, Kent.
Editorial office: 47 Bedford Square, London WC1B 3DP

Photoset by Rowland Phototypesetting Ltd
Bury St Edmunds, Suffolk
Printed in Great Britain by
Cox & Wyman Ltd, Reading

**British Library C.I.P.**

Baxter, John
  Who burned Australia?
  1. Forest Fires–Australia–History–20th Century
  I. Title
  994.06'3      SD421.34.A8

ISBN 0-450-05749-6

# Contents

## Acknowledgments

Among the references consulted in compiling this account were the following:

*Cockatoo. Ash Wednesday, 1983. The People's Story.* Edward Mundie. Hyland House, Melbourne, 1983.

'Ash Wednesday'. *Age* newspaper/*Adelaide Advertiser*, 1983.

*Great Australian Disasters.* Ian Mackay. Rigby, 1982.

*Bushfires: Their Effect On Australian Life and Landscape.* Ed. Peter Stanbury. University of Sydney, 1981.

*The Use of Fire in the Forest Environment.* R. R. Richmond. Forestry Commission of New South Wales, 1976.

I am also extremely grateful to Michael Caulfield, Lee Harding, Virginia Baxter and the librarians of the *Australian*, particularly Ms Marie Lloyd, the *Sydney Morning Herald*, the *Adelaide Advertiser*, the *Age*, *Hobby Farmer* and other periodicals whose reportage of Ash Wednesday has been consulted or quoted.

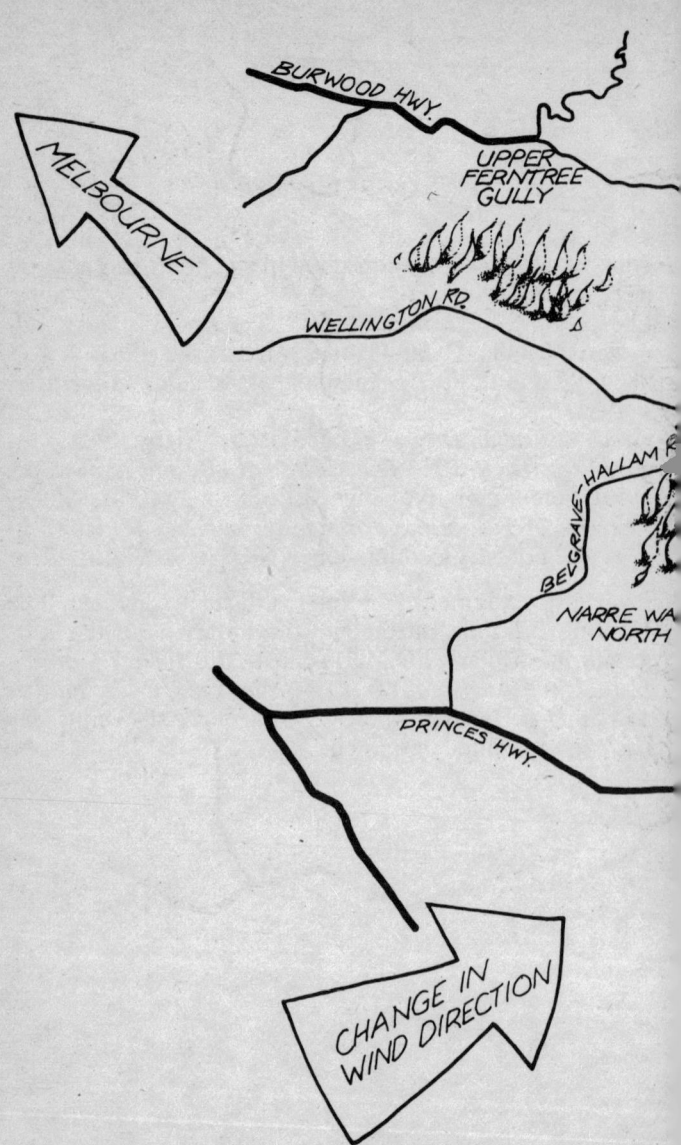

Map of Dandenongs Fires, Ash Wednesday, 1983

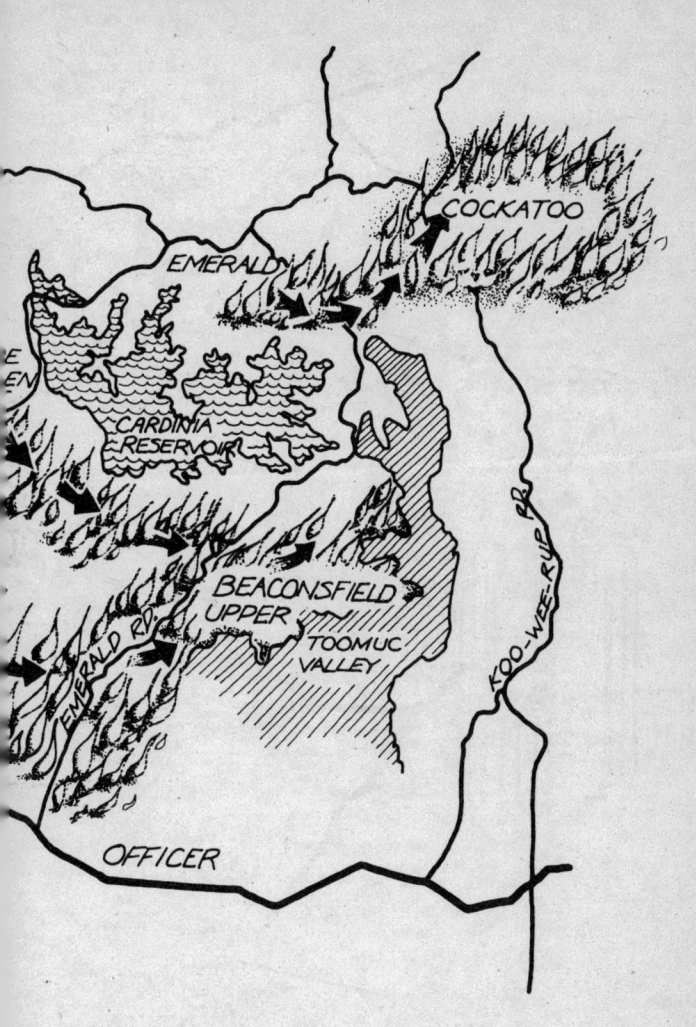

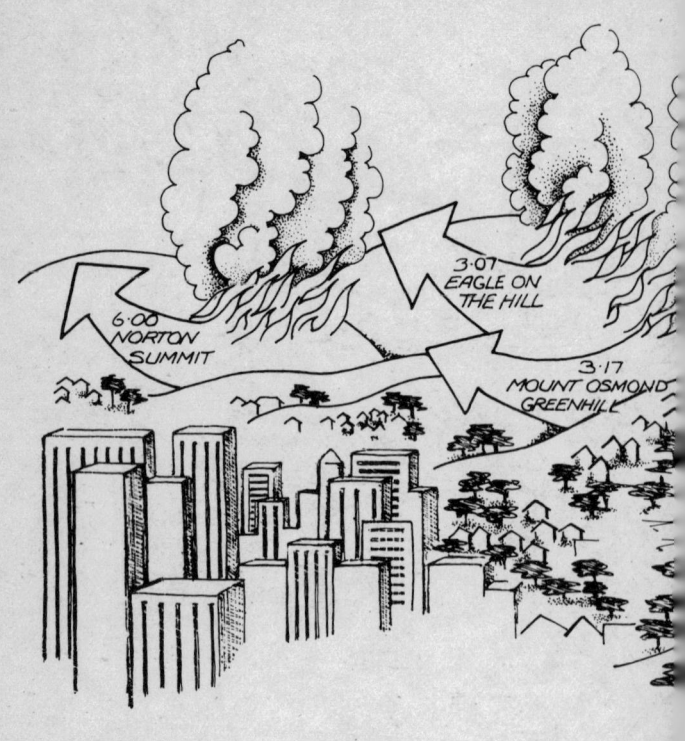

Map of Adelaide Hills Fires, Ash Wednesday, 1983

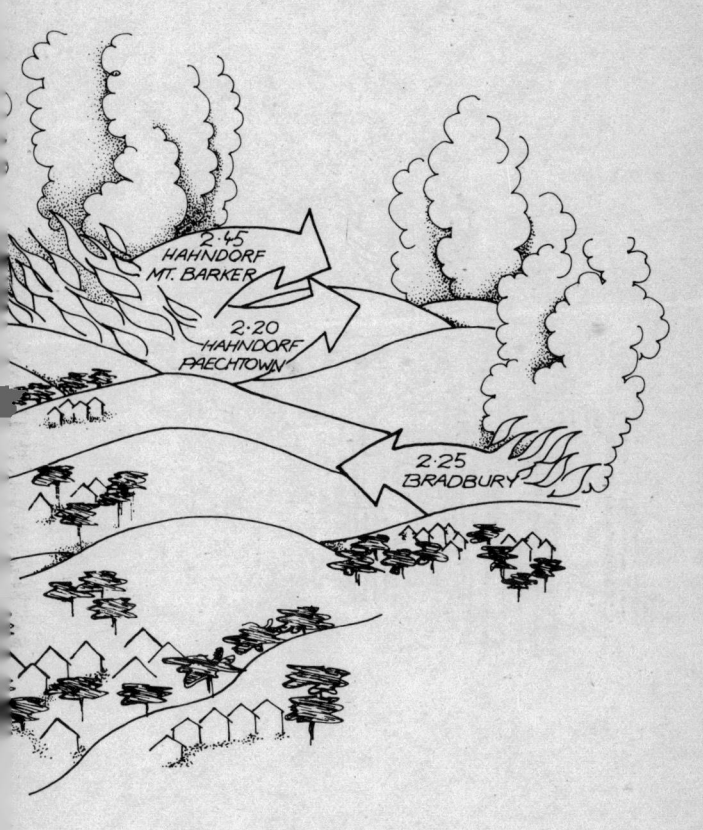

## Introduction

IN TWO weeks of February 1983, the last month of summer and the first week of March, the Australian continent suffered an ecological disaster unprecedented in its history and scarcely rivalled in the global record.

Seventy-one people died. Many more were maimed. The cost of homes and stock destroyed can scarcely be calculated but one estimate places it at $400 million. The loss in environmental terms – forest and woodland, non-domestic animals – is beyond evaluation.

Although the fires of 16 February, Ash Wednesday, in southern Australia attracted most media coverage, they were not in themselves the whole disaster, nor even its greatest part. The Australian environment was like an old mansion, run-down and vandalised, its foundations sinking, its roof gaping to the elements. Then someone dropped a match in a pile of debris by the door . . .

The fires of Ash Wednesday merely brought to its horrifying climax a process of disaster which had been prepared by nature for at least two years and ended three months later in an unexpected and ironic climax.

Unravelling the story of Ash Wednesday is like an exercise in forensic investigation. We enter the roofless ruin of the house, still stinking with the water poured on by helpless firemen as they tried to quench the runaway flames. It seems a simple enough case of neglect and decay culminating in an accidental blaze.

And yet . . .

There are witnesses who step forward to insist the old house was intentionally vandalized, for profit. That the fire was set. We take out our notebooks and begin to reconstruct a mystery. Or perhaps seventy-one crimes.

# ONE

## *'Like all Australians, I thought I knew about bushfires . . .'*

THE FIRES of Ash Wednesday 1983, which swept through south-eastern Victoria and the Adelaide Hills of South Australia were the worst fires the world had ever seen.

But they were only part of a nationwide disaster which in turn related to even more catastrophic events taking place in South America, South Africa and on islands across the Pacific. The sudden disappearance of sealions from the Galapagos Islands and birds from the guano islands off Peru, a season of floods in Ecuador and a massive redirection of the Humboldt Current all played a role in a gigantic ecological opera.

In Australia the damage was triggered and exacerbated by a number of events, some random, some calculated.

The Ash Wedensday fires were predictable. They were avoidable. Some were also deliberately lit.

Most people in Australia have a bushfire story. In a country still mostly empty of people, fire is a risk faced not only by farmers and country-dwellers but by every bushwalker and weekend picknicker.

The man who tells the following story is now a major film director. In the mid-1970s as a university student he spent his vacation on a grazing property in north-western New South Wales owned by the father of a friend.

In the middle of a long drought period fire broke out. He

and a number of other teenagers were co-opted on to the 'wet sack team' battling small outbreaks on the fringe of the blaze. As the fire began to run out of control, an open-backed utility truck – the ubiquitous 'ute', Australia's standard outback transport – arrived and took them to the main firefront, deep in the bush:

It was unlike anything I'd seen before. I'd seen bushes and grassland alight, but never trees 50 foot tall with flame from the bottom to the top.

That's when the noise started to hit me. I began to hear explosions.

We were on a 'fire track' – a dirt road gouged out by a bulldozer. On both sides of us, trees were blowing up – literally.

You wouldn't know a particular tree was going. You'd hear a 'whoomph'. Then it would rain down on you. Bits of twigs. Bits of burning leaves.

You would see a 'ball of fire' – not spherical; a co-agulated mass of flame, deeply orange in colour – move across 20 feet of ground; just jump from one tree to another.

It was then that it jumped right over the top of us.

We all went flat to the ground. Not by choice, but because you were literally forced, face down, into the dirt.

You know people, talking about fires, say they're like living animals? There is no better metaphor. It does seem like some monster that has design and will, and isn't just 'raging out of control', in the typical newscaster's words, but literally is out to get *you*, and everything around you.

It was intensely personal. I wasn't even aware of people around. It was like this thing was out to get me.

I couldn't breathe. I simply couldn't find air. I was like an asthmatic. Every time I took in breath, it hurt. It hurt so much. It was like someone was gouging the inside of your throat. I thought we were dead.

16

One of the older guys got to his feet and all he said was, 'Run!'

He meant back the way we'd come.

Like all Australians, I thought I knew about bushfires. I didn't.

It's useless to think of a fire as a straight line. It spits, and comes back and spits again until it gets a hold on the next part. Then it seems to roar ahead.

You'd run. Then, literally, a tongue of flame would shoot out, then pull back again if the bush didn't catch fire. Then it would jump forward at another point.

We were running through an area that was partially burned or burned out. I had no idea where we were running to and I don't think anyone else did either. If the fire had been headed in our direction, none of us would have been around to tell the story. It was just too quick. Even in a vehicle, you'd be lucky to outrun it on those circuitous bush roads.

We ran for what seemed hours. Some were running and crying. Guys would stumble, fall down in the dirt. Nobody stopped to help. The standard things happened, I found later. People pissing in their pants and so on.

Everybody looked incredibly different. Apart from the terror implanted in our faces, some of us were covered in ash. I had a beard and it was singed. Others had lost eyebrows.

We came to a small tributary of the creek that ran through the property. The stream was about 12 feet wide, maybe 3 feet deep in the deepest pool. And I thought, Ah, we're safe. One of my friends screamed, 'Stay here, stay here.'

But the old guy who led the rush just went straight through. The other men didn't even bother to tell us it was crazy. They kept going.

Then we were out of the bush and into the paddock area, all of which had already been burned and in some places was still burning.

The wind was roaring. I saw a big hay shed. Just a roof of corrugated iron supported on huge steel stanchions and hay stacked up to the roof. It was totally alight.

As we came close to it I saw the roof begin to buckle and lift in the wind. Corrugated iron doesn't melt, but the steel uprights did. They dripped. Literally dripped. It was as if the whole thing just dribbled into the ground. The iron sheets would buckle, then just fly – take off into the air.

This fire destroyed hundreds of acres of farmland, miles of valuable fences and stock valued at hundreds of thousands of dollars. Two men died in the blaze.

By Australian standards it was less than a major fire. By comparison with Ash Wednesday it is not worth considering.

# TWO

## *An ancient evil*

BUSHFIRES ARE a constant of the Australian environment, as predictable as the Russian winter or the Indonesian monsoon. Fifteen thousand of them are recorded each year.

Natural causes like lightning and occasionally spontaneous combustion sparked bushfires almost as soon as the climate turned dry enough to render the forest inflammable. The dried bed of Lake George near Canberra yields sediments with charcoal fragments scattered through the geological record for the last 700,000 years.

Man had not lived on the continent for long before fire became a major feature of the environment. The aboriginals lit fires and kept on lighting them. Most would have been accidental; a camp fire left untended, a spark falling from the 'fire stick' with which the tribe carried coals during the daily search for food. But often the early men of Australia calculatedly set the bush on fire.

For all the apparent crudity of his existence, the tribal aboriginal was in fact a shrewd exploiter of the environment, husbanding its resources with an economy never approached by the white farmer or grazier. When he fired the bush it was to achieve specific results.

The explorer Ernest Giles wrote of the aboriginals, 'One would think they lived on fire instead of water.' The naturalist Vincent Serventy has detailed the aboriginal use of fire. 'They used it for signalling so that all would know

19

where a group was hunting, and to clear the ground which made it easier to hunt in the forest. Fire was often used to drive animals into a hunting ambush. Such fires also promoted the growth of palatable young shoots, the "green pick" which brought game into the area.'

Serventy also quotes an 1848 comment by the explorer James Mitchell. 'In summer, the burning of the long grass also discloses vermin, birds' nests, etc., on which the females and the children who chiefly burn the grass, feed.' Mitchell went on, 'Fire, grass, kangaroos and human inhabitants seem all dependent on each other for existence in Australia.'

Mitchell was right of course. The aboriginal saw the land as a living entity, a provider which would feed, clothe and shelter him if he placated its spirits and respected the natural order.

Each season produced its bounty. Grass seeds for pounding into cakes, fruit and bulbs in the spring, roots in the winter. Grubs clung to tubers underground and hid under the bark of rotting trees. Game was killed when plentiful. As it died off or moved away the tribe moved with it, knowing that in its absence the land would renew and replenish itself.

Fire had its place in this cycle. But aboriginals knew better than to risk setting the forest ablaze, damaging the trees whose long, straight trunks and limbs provided them with wood for spears and bark for clothing. The ideal at which they aimed was a park-like, open woodland, neat and easy.

White settlers knew only one use for the land: intensive farming and grazing. In preparing the ground for that use fire was a simple, if inexact tool. 'Slash and burn' is a crude system of proven short-term efficiency. Trees and brush were torn out or severed just above ground and the foliage burned, its minerals turned to ash and fed back into the ground.

Grass would flourish in such ground for a season or two.

But farmers ignorant of soil chemistry did not realise the waste products of a eucalyptus burn included volatile substances which repelled water. If these found their way into the soil they established a waterproof layer a few inches below the surface. Water ran off such layers, causing erosion after the first bad rains. Imported European grazing animals, in particular the sheep which Vincent Serventy, in a well-turned phrase, christened 'walking vacuum cleaners', quickly turned savannah to close-cropped lawn, then to naked earth. Erosion cut deep gullies into the hillsides and rain washed valuable topsoil into the rivers, which in turn silted up, creating ideal conditions for flooding.

Of the aboriginal's shrewdly self-interested respect for the environment, there was little sign. Most Europeans found the continent ugly, even repellent. The surgeon of the First Fleet, John White, called it, 'A country and place so forbidding and hateful as only to merit execration and curses.' Anthony Trollope in 1876 remarked that: 'It is taken for granted that Australia is ugly.'

The editors of *The Last of Lands,* an influential environmental anthology, summed up the prevailing view of those early settlers: 'The gibber plains and sandy gorges, leathery-leaved eucalypts filled with screeching birds, and the eerie, monotonous bush were usually considered repulsive by colonists fresh from the hedged landscape of the pretty English shires.'

Such attempts as were made at environmental improvement were uniformly disastrous. Homesick settlers imported both the rabbit and the sparrow, the first becoming a plague so extreme that only another introduced evil, the disease myxomatosis, could bring it under control, the second flourishing but changing out of all recognition.

For the Australian farmer fire remained a legitimate tool of cultivation. 'Burning off' is a common form of clearance in rural areas. A paddock or domestic backyard will be set on fire as routinely as a British farmer mows or cultivates a similar piece of land. Stubble is burned in the time-honoured way to pass nutrients back into the land.

It was inevitable that fire would get out of control under such conditions. The bushfire became a summer fixture, as immutable as the opening of the cricket season. Every summer the bush burns, especially where man interfaces with it. Around the big cities the scenic and heavily timbered residential areas burn. The national parks burn. The farmlands of Victoria burn.

The 1851 fires were succeeded by the 1897 fires in Victoria's South Gippsland. In 1926, half a million acres in Victoria were burned. In January 1939 the fires that earned the still-traditional title of 'Black Friday' killed seventy-one people.

In 1967 Tasmania, always regarded as free of fire risk because of its well-watered fertility, exploded in bushfires after seventy years of peace. In four hours sixty-two people died. Victoria remained the state most under threat but in 1968 fires destroyed hundreds of acres of forest and many homes in the Blue Mountains outside Sydney in New South Wales. And ironically on Ash Wednesday 1983 South Australia suffered some of the worst fires in its history.

The public attitude to this recurring plague is a puzzling indifference despite the fact that after each fire there is indignation, horror, generosity to the survivors and ringing assurances that, 'This will not happen again.'

In the wake of the 1968 Blue Mountains' fires a conference between Commonwealth, state and council members came up with a wide-ranging plan to ensure that fire would never again destroy this green and valuable district. A survey was planned to isolate the major causes of bushfire. Particularly inflammable varieties of tree would be identified and removed. A woodchip plant was suggested as an economical, even profitable means of using up these inconvenient natives of the bush. Residents with homes in fire danger areas would be offered another site in exchange. Conferences would be held on more efficient firefighting and the co-ordination of local resources. It was suggested that Commonwealth firefighting funds be channelled through the local councils, who knew best where to spend the money.

It may have been this politically sensitive suggestion which doomed the enterprise. It is more likely that the political value of fire prevention waned with the fires themselves. Five years later there is no woodchip plant, little reafforestation, even less co-ordination. Fighting fires is still a job given to the dedicated amateur driven by the realisation that, while he saves one man's house another volunteer brigade may be saving his. And most people continue to think that it probably won't happen again.

That Australia is a continent as peculiarly inappropriate to the use of fire as the steppes of Russia are to citrus-growing has yet to be grasped by the average grazier. One of the people who contributed his experiences of bushfires to this book remarked:

'All Australian graziers I've met are European in their attitude to the bush. Their attitude is, "Don't work with it. Work against it. You cut down the trees and the suckers will come back and get ya." Not just come back – come back and "get ya". They'll do it deliberately. If it rains, you blame it on someone.'

The bushfire represents a logical extension of this attitude. The risk is always there. The farmer has created it, first by farming in an environment unsuited to such land use and second by exposing an incendiary landscape to the constant threat of ignition. For all the horror of the bushfire, it is seldom actually a surprise to the farmer.

The fires of Ash Wednesday were more than a surprise. They were a shock, visceral and tragic. They were not, either in their form or their scope, comparable to the fires of earlier years. One authority estimated that the power of the one which destroyed the Dandenongs outside Melbourne was thirty times that of a conventional Australian bushfire.

The fires were merely one aspect of an environmental assault unprecedented in the history of the Pacific – an assault that continued for weeks after the fires, when it took a new and totally unexpected form. It is an assault that is continuing still.

# THREE

## *Rent-a-bitey*

A CHILD in rural Australia grows up with the conviction that everything is out to get him.

Though mostly paranoia the belief does have its basis in fact. Partly through his own efforts the white Australian has created an ecology hostile to his life there. Sometimes by chance, mostly by design, the Australian places himself in opposition to his environment rather than in tune with it.

The result is a sustained tension between man and nature. One learns early in life to avoid any cover that might harbour one of Australia's various poisonous snakes, insects or spiders. Tall, dry grass, piles of firewood or the dark spaces under buildings (themselves dictated by the need to lift wooden construction above the attacks of termites or 'white ants') are all an anathema to the wary countryman.

No family with an outdoor toilet ever forgets to raise the seat before use, since the warm darkness underneath it is the favoured hiding place of the red-back spider, whose bite is highly venomous. The funnel-web, equally poisonous and trapdoor spiders, lurk in tunnels bored into soft earth.

A Sydney security firm called 'Rent-a-Bitey' supplies deadly reptiles, insects and spiders as living guards for jewellery displays. Its team of natural sentinels includes centipedes, scorpions, Moray eels, the blue-ringed octopus, stone and puffer fish, the spiny butterfly cod, as well as the funnel-web spider, a star at $200 a week.

The Australian climate encourages the growth of plants and animals more suited to semi-desert landscapes. It is no coincidence that the three areas of the world most ravaged by bushfires are south-eastern Australia, California and the Mediterranean coast of France, all of which share a similar climate, plant life and patterns of settlement.

Australian gardens are filled with adapted species from these countries: the spiked yucca, common in California; pampas grass, a saw-edged native of the Camargue. Cacti flourish. The prickly pear spread in plague proportions across desert Australia until ecologists imported the beetle *Cactoblastas cactorum* to wipe it out.

Before white settlement the Australian landscape looked very different. 'Nothing grew in profusion,' wrote the conservationist Eric Rolls. 'The flowering shrubs grew singly or in little clumps. The grasses, many of them perennial, grew in spaced tussocks. Many other plants grew among them but there was seldom complete ground cover.'

Kangaroos and other native animals nibbled with sharp teeth, leaving grass roots intact, the natural ground cover untrampled. Sheep and cattle altered that. As Rolls records:

In most districts sheep changed the soil forever in about six years. Hardened ground ringed out from the yards. Bruised plants died. Ground never stays bare for long. Inferior Australian grasses with cruel spiked seeds took over from the good grasses with millet-like seeds. And European weeds imported in vegetable seed, in woolly sheep, in horses' tails, in chaff sprang up and thrived with no competition. They had had thousands of years' experience of hard ground and cloven hooves.

It often seems to the visitor that every Australian plant has the capacity to scratch, puncture or sting the unwary.

On moist ground nettles flourish. The dry earth produces hard, spiky grasses, and a flat, deep-rooted bush called the bindi-i, with star-like thorns that can puncture a bicycle tyre.

Waves of ecological interference have created a forest environment alien to European conceptions of woodland. The standard 'bush', covering the drier areas of the coast, most of the mountains and large portions of the non-desert interior, consists of stringy, rough, apparently barkless eucalypts, from whose tufted tops dangle the long, tough leaves so frustrating to eighteenth-century artists, accustomed only to leaves which sprouted in all directions from a wide canopy of branches.

Set wide apart, the trees climb with an often eloquent sinuousness toward the light. There are seldom any branches below 10 feet and these reach upwards with the same eagerness as the trunk. Scattered around most trees are 'suckers'; saplings seeded from the main tree. They compete for space with forms of brush and flowering plants as tough and hardy as the trees themselves.

The ground underfoot is usually hard, dry and impacted – a contrast to that visualised by Eric Rolls as prevailing in pre-settlement times: 'No boot had trampled our soil, no wheel marked it. No shovel had ever turned it. There was a mulch of thousands of years. The soil itself seemed to be growing. The top five centimetres could be raked through the fingers. The subsoil was friable and rich with decomposed roots.'

Such a forest floor must be almost unknown in contemporary Australia. Scraps of bark, shed by the trees or torn off by weather, litter the ground intermixed with leaves and twigs, still apparently alive, the moisture conserved inside the tough, leathery foliage. Being evergreens the gums shed their leaves not merely in autumn but throughout the year, quickly layering the ground with fuel for bushfires. Among the twigs are seeds, hard as nuts, or drying flower heads from bushes like the banksia, as bristly as a bottle brush, the spikes protecting a woody cob studded with the knobs of seed pods.

City visitors to the bush tend to class any tree as a 'gum' – a term coined by early settlers because many of the trees exude a thick, resinous sap which hardens like amber on contact with the air. Gums or eucalypts make up only a proportion of the bush trees, however. Acacias, including the vivid yellow wattle, the national emblem, are common. So are ti-trees, paperbarks and sub-varieties of the eucalypt like scribbly gums, stringybarks and ghost gums and the so-called slaty box and Blue Mountain ash. Among the shrubs banksia, waratah, Christmas bush and the spiky blackboy flourish. Most of these, especially the vividly crimson waratah, have either waxy, water-conserving leaves and flowers or bristling pods to protect the seeds against anything the bush can throw at them.

The Australian bush is tough; it contains perhaps the toughest flora in the world. It is not conventionally beautiful. What it does best is survive. The next best thing it does is burn.

As long as rain falls occasionally, the ground cover burns poorly. At a temperature of 15°C and with a humidity of 10°C a fire will barely sustain itself. Raise the temperature to 35°C; however, and flames 5 feet high will spread at 30 feet an hour. Should temperatures climb to 40°C – far from rare in the Australian summer – and humidity drop to 5°C, the ground cover becomes tinder-dry. Fan a fire with a 30-knot wind and, to quote the Forests Commission:

The resulting fire will be characterised by a rate of spread, disregarding spot fires, of 3 kilometres [2 miles] an hour, a flame height of 50 metres [33 feet] occurring in explosive surges, and a heat output of such magnitude that every 200 metres [132 feet] of fire edge could provide more than enough power to meet the peak electricity load of Sydney and Melbourne combined.

Bushfires preceded man to Australian by millennia.

Even the lush rain forest of the moister, more tropical Australia of 100,000 years ago burned on occasion. When the weather entered a drier cycle 70,000 years ago plants were already evolving fire-resistant and fire-dependent strains. By the time man arrived, 40,000 years ago, fires were already a feature of life. Started by lightning, they tore periodically through the savannah and forest, destroying those species not adapted to their searing heat.

The drier climate had already driven back to the coast those plants able to survive only in areas of high humidity and rainfall. What remained were the tough-skinned water-hoarders with deep root systems and thin foliage. To conserve the meagre water some converted it to an oil which fed fluid to the narrow, leathery leaves.

The now-inflammable bush plants evolved to survive fires. Rough-barked varieties did not live long; their exteriors led the flames straight to the leaf canopy. The eucalypts flourished however. Gum bark contains a substance called kino, a fire-inhibitor. Some species of tree developed lignotubers, swellings where the trunk met the earth. These gnarled thickenings absorbed the worst heat of a grass fire leaving the trunk undamaged.

Other plants incorporated heat into their breeding cycle. Only a fire can crack the seed pods of the inflammable banksia. Many gums carry epicormic buds beneath their bark. Exposed when a fire strips off the outer layer, they sprout independently of the main root system, cloaking apparently ruined trees in new growth.

An ecologist has described the bush's response to fire in all its contradictory unexpectedness:

A wildfire sweeps through the bush, leaving the landscape blackened, bare and seemingly lifeless. But life soon returns. Burrowing animals – spiders, insects, reptiles, mammals – emerge from their underground refuges. Others, which escaped the fire by moving out of its path, on the ground or in the air, return.

New shoots appear on the branches and trunk of trees, around the bases of trees and shrubs, and from underground roots,rhizomes and bulbs. Seeds, buried in the soil or newly released from woody capsules, germinate. Within weeks, if there is sufficient moisture in the soil, the burned area is a mass of new growth. Some years later, there are few signs of the original devastation.

To the visitor the Australian bush on a hot day can seem to be waiting for fire. The gum leaves droop with oil. It vaporises in the heat, filling the air with a faint, blue-grey haze that softens outlines as close as half a mile away. Shreds of bark trail from the paperbarks and Blue Mountain ash, dangling like fuses into the accumulated ground cover, tinder-dry on the hard ground. Fed from deep roots the plants are alive but moist only in their cores. The outer layers are tough and dry.

This ecology has evolved to endure a fire, then climb back to life. In this, it contrasts with the artifacts man has scattered through the forest. A gum tree will sprout again, a banksia reproduce itself through its seeds. But cars, houses, and man himself have no protection from fire, no mechanism of adaptation. The fire that wounds a forest leaves man and his works dead, incinerated as if they never existed.

# FOUR

## *Run for your life*

BUSHMEN WILL tell you that animals know when fire is about. No less than the bush itself, its animals have adapted to the constant risk of fire and developed strategies to escape it. The first is to run.

Blind panic plays some part in the animals' urge to run from a fire, but there is an ingrained sense of self-preservation there too. Bushfires are essentially fronts of fire moving swiftly through the environment, driven by their own self-created draught and by the devouring need for more and more fuel.

Trees left behind by the fire will continue to burn, sometimes until nothing is left but a glowing pit in the ground, but more often they will smoulder for a time, then go out. Once burned ground cover offers no continuing fuel for a fire, so only a few minutes after a fire the ground itself is cool enough to stand on.

Birds survive best of all. They merely fly above the fire and settle on the far side. Unless a fire is so voracious as to burn all the available oxygen, they run little risk. They have even been observed hovering around the edge of a fire, picking up insects driven out of the flames.

Ground-dwellers retreat to their burrows. The wombat seldom suffers in a fire. Rocks harbour frogs, lizards and small snakes; the fire moves so quickly that their moist hiding places are seldom even made uncomfortable.

Aquatic creatures seldom survive though. In drought

31

streams dry up long before the fire season begins. Any fish left alive will be boiled as the fire sweeps over their shallow habitat. Eels have been seen after a fire with 'their whitened, bloated corpses floating in the tea-dark water'.

A woman returning to her burned-out house in the town of Greenhill in South Australia after Ash Wednesday found pathetic evidence of the fire's strength: 'The tortoise was half in, half out of his pond, cooked. That upset me most of all.'

The urge for humans to lie down in a stream and rely on its protection is almost always fatal, since the water quickly boils and the resulting superheated steam proves more dangerous than the fire itself.

The worst fatalities among fauna are suffered by leaf-eating or sap-sucking tree creatures like the koala and the opossum. Slow-moving and arboreal they have no place to hide and the aftermath of fires reveals hundreds of their charred bodies.

Running remains for most animals the common and sensible response and they will usually suspend their normal animosity towards each other to flee from a fire. One bushfire survivor commented, 'You've never seen so many odd pairings in your life. Snakes wriggling beside marsupial mice; lizards, feral cats, a few goats. Echidnas.'

The majority of animals in the wild survive fire. Biologist Per Christensen radio-tagged thirty animals during an experimental burn in Western Australia. Only one died, a woylie which took refuge in a log that caught fire. It suffocated.

Unlike wild animals, domestic pets and farm stock lack all instinct for dealing with fire. Sheep in particular mill around in total panic, usually congregating in a dense flock and hence vulnerable to the greatest risk of any fire, oxygen starvation. After a fire whole flocks of sheep can be found in the corner of a paddock, dead from asphyxiation but totally unmarked by the flames. Horses become hysterical. Cattle blunder into fences, become caught up and suffer such burns that they inevitably need to be shot subsequently.

Ironically the national parks movement of recent times has placed wild animals at increased risk from fire. Frequently parks and reserves exist as islands within a suburban environment. Caught between a firefront and a freeway or housing development animals often burn in their terror and confusion.

The effect of fire in an enclosed situation has been graphically described by Harry F. Recher of the Australian Museum's Department of Environmental Studies:

During the Hobart fires of 1967, large numbers of birds and mammals were killed. It was the same at Nadgee in south-eastern New South Wales when a wildfire burnt the Nature Reserve in 1972. Kangaroos were seen fleeing from the flames and hurtling over cliffs into the ocean. Birds were swept up in a fire storm, suffocated and dropped into the ocean. Later they washed upon the shore in windrows like so much seaweed. Possums were baked as they clung to branches or huddled miserably in the middle of dirt tracts.

After Ash Wednesday people recalled the response of animals to the risk of fire. Dogs whined and prowled restlessly. Birds, affected by the heat, flew randomly, without their usual tendency to gather in small flocks. When they settled it was to perch with heads cocked, listening and watching. Fire unites all animals – from the fires of Ash Wednesday there would be, for many of them, no escape.

# FIVE

## *'God, where did all the water go?'*

BY THE middle of 1982 the whole of Australia knew something had gone wrong with the weather.

Rainfall is always meagre in Australia except in the mountain-guarded strip between the eastern coastal ranges and the Pacific, but for almost a year it had barely rained at all.

By the end of June 1982 [noted *New Scientist*], serious or severe deficiencies in rainfall had developed over most of New South Wales, central and south-western Queensland and northern Victoria. By the end of August rainfall deficiencies of three to five months duration had become established over much of South Australia, almost all of Victoria and New South Wales, Tasmania and southern Queensland.

The failure of winter rains was followed by abnormally small rainfall in spring and by the end of 1982 the severe conditions extended over most of eastern and southern Australia.

Behind this terse description lay a drama which Australians could see played out around them throughout the last months of 1982. For country people drought had been a fact of life since the previous summer. The rivers had long since dried up. The creeks were empty or merely disconnected strings of pools. In many areas even the artesian wells, their pumps turned by the ubiquitous country windmill, had dried up.

Now city-dwellers felt it too.

The dams that served Sydney and Melbourne began to dry up. By February 1983 Eildon Weir near Melbourne had sunk 80 feet. 'My first reaction was "God, where did all the water go?" ' wrote photographer Mark Ashkenasy as he covered the familiar lake for the Melbourne *Sun*. 'The bed is cracked beyond belief. Some of the cracks go down eight inches.' A 60-foot tree whose upper branches had once been a hazard to water-skiers now stood fully exposed in a desert of crazed, drying clay.

Burrinjuck Dam, which serves Sydney, dwindled to only 3.5 per cent of normal capacity. A railway line built in 1908 to cart sand for the building of the dam and under water ever since, raised rusted rails and rotting sleepers above the surface. Tap water was often murky, faintly tainted in taste.

Suburban gardeners complained of restrictions on watering. A few hours with a hand-held hose each week barely kept alive the lawns and garden beds that, for many fiercely territorial Australians, are the greatest pleasure in life.

Patsy Adam-Smith in the *National Times* described hard times on the suburban front.

Water restrictions are severe in the city. Gardens can be watered by hose only three nights a week from 7 to 9 o'clock. Few guests turn up for dinner before 8.30 on these nights. Lawns are not permitted to be watered by either hose or bucket and at a backyard barbecue two weeks ago I saw a table fork disappear down a five cm fissure in what used to be the lawn. Swimming pools may be filled only by bucket on the three nights specified. Cars may be washed only by buckets.

Writing in the *Australian* Lisa Kelly sounded the indignant complaint of the environmentalist. 'We're a spoiled society,' she wrote 'wasting water on our personal needs

36

and lamenting that we can no longer pour litres and litres of this life-saving fluid onto our pampered gardens.'

The government issued bumper stickers: WATERHOLICS ARE RUNNING US DRY. A few people were prosecuted for watering out of the approved hours. It did no good. Hints for more provident use of water, including placing a house brick in the toilet cistern to reduce the amount of water flushed away, were also ignored.

Melbourne is a city in the British tradition, a sub-tropical Birmingham set on flat country some miles from the coast, but ringed to the south-east by wooded hills that contain both the city's most luxurious suburbs and hundreds of cheap weekend retreats, hand-made houses of pine and glass thrown up by cheerful amateurs on half-cleared hill-side blocks. Melbourne weather is usually cool and rainy. Fires in the inner suburbs could find little on which to feed either in the trimmed lawns or low sandstone buildings.

Sydney, despite its location along the rambling shores of the harbour, contains enough scrubland and forest to make fire a continuing summer problem. The areas of protected woodland around the city ignite routinely every year. So do the Blue Mountains, 50 miles west of the city, whose narrow, forested valleys echo those in the Dandenongs outside Melbourne and the hills behind Adelaide.

In January of 1983 even the people of the cities recognised that something extraordinary was taking place in their country.

People joked about an Australian film, *The Last Wave*, made by Peter Weir five years before. A precognitive lawyer visualised Australia overwhelmed by a new flood. 'Hasn't the weather been strange lately?' enquired the advertising copy. 'Perhaps it could be . . . a warning?' 'Perhaps,' some people laughed.

Even in the moist coastal strip, the weather was bizarre. Driving through the winding Blue Mountains' roads one would abruptly find the road dissolving in a cold grey fog. The large homes, set in forests back from the road, the

Edwardian resort hotels, even the thick bush itself all faded in the clinging mist. Christmas Day 1982 was, like that of the year before, overcast and cool, even though we were in the height of summer. In the afternoon it rained.

A pattern was established.

By day the air was filled with a fine talcum-like dust. Too thin to be more than a slight irritant to the eyes, it became more visible in the mornings when one found one's car filmed with a rusty powder. Water from the windscreen washers flushed away a dull rose.

As the days became hotter Sydney began to smell. The water in the harbour bays was sluggish, coated with rubbish not swept out in the morning tide.

Life continued, luxurious and easy. King prawns were six dollars a pound and plentiful. Australians delight in their seafood. At 'Prawn Nights', members of the large working mens' clubs in the Sydney suburbs gorged themselves on mountains of pink shellfish and gallons of cold lager. The smell of the rotting shells seemed to be everywhere, disturbingly in harmony with the fruity sweetness of the white-flowered frangipani.

The heat and humidity of Sydney could be intolerable. Only the promise of an evening change gave one hope. This, too, is part of the pattern of Australian coastal weather – the thick, stifling day, then the hard, cold slap of a squall.

Sydney has the 'Southerly Buster', phenomenon of the spring and summer for which eastern Australians pray at the end of a baking hot day. Roaring out of the south at gusts of 40 knots or more and driving before it a wall of cold air 20°C below that cloaking the exhausted city, the Buster rattles blinds, flaps curtains, whips sand into the faces of sunbathers and tips yachts in the harbour on to their sides.

The Southerly Buster is a familiar feature of the weather, which in Australia is characterised by sudden and dramatic changes, especially in temperature and wind velocity. But, like the drought, nobody knows why it happens. In the summer of 1982 the Australian Numerical Research Centre was constructing a mathematical model of the Buster from Landsat photographs, readings of temperature, pressure and wind speed, but they could neither explain the phenomenon nor predict it.

Those who read history might have foretold what was to happen next. But Australians generally decline to consider history. Patsy Adam-Smith, herself trapped in the 1962 Tasmanian fires which killed sixty-two people, saw the whole process with the hindsight of three generations who suffered burn-out.

> People will keep saying that history repeats itself. Well, of course it does. That's the whole point. If people learnt from history then history would not repeat itself. If an author writes, hoping that by telling the story it may help mankind to understand an event, a man, a place in time, last week's fires must drive him to despair. Sometimes, while reading or televiewing the news as the death toll mounted, a body could be forgiven for crying. 'I've seen it all before!'

Nature even offered a harbinger, a sign of what might be in store for Melbourne and its satellite towns if plans were not made immediately to fend off fire.

Tuesday, February 8th – a week before Ash Wednesday – was Melbourne's hottest February day since 1901:43.2°C at two thirty-five in the afternoon.

Then, at 3 p.m., a front of cool air rushed out from the west, driving before it a wave of dust scooped off the parched and defoliated plains on which no rain has fallen for almost a year.

The storm rolled in a curtain 350 miles wide and 3 miles high. A thousand square miles of the Victorian landscape was blotted out. An amazing 150,000 tons of soil, organic matter and nutrients – the topsoil of countless farms – swept across Melbourne on its way to the sea.

The people of the city experienced it as a reddish–brown dust storm, impenetrable as a wall, which filled the sky – a phenomenon from the Sahara rather than this temperate, green coastal area.

Suddenly Melbourne's wide streets were lost in a red haze. Headlights and street lights went on in mid-afternoon. All three airports were closed. Kim Lockwood's report in the *Australian* described: 'City workers huddled in office doorways talking of the end of the world as the dust clogged their mouths and noses and seared their eyes.'

Sweeping over the city it bore down on the beach resorts along Port Phillip Bay and the coast, cloaking the pretty weekend cottages at Airey's Inlet and other towns along the Great Ocean Road. Parents sunning themselves saw the cloud coming but before they could reach the water to rescue their children visibility was reduced to no more than a few yards.

*Could* it be a warning?

# SIX

## *The drought and the Christ Child*

THE DUST storms, the drought, the unseasonal changes in temperature and rainfall along the coast were warnings, of a kind. But few people in Australia could read the signs.

Australians adopt a phlegmatic resignation towards the weather. It is the nature of the elements, they feel, to trick and betray them. Why prepare for drought when the almost inevitable result will be floods that sweep away your carefully constructed dams and rot your stored fodder? And why put in culverts to divert flood waters when the creekbeds will almost certainly remain dry for a year?

Australian scientists have not been deterred from experimenting on what is, in many respects, a perfect laboratory for climatological research. The continent sits on the border between a number of weather systems. To the south lies the ice mass of Antartica and the band of east blowing gales, the Roaring Forties, which can push icy air masses up into southern Australia. Centred over Indonesia, northwest of Australia, is a low-pressure system of warm, moist air whose effects render the area among the most humid and debilitating on earth. Balanced between these two extremes, weather in western and central Australia tends to remain static, a harsh, dry, semi-desert environment.

The east coast, however, faces the wide Pacific. This coast of Queensland, New South Wales and Victoria is an international playground. Between Bondi Beach, outside Sydney, and the coast of Argentina lies the Pacific, its surface heated by the sun, its prevailing westerly winds funnelling warm, water-laden breezes across to the Australian coast. The winds air-condition the beaches, water the farmlands of the coast, dump rain on the mountains and carry the rest of their moisture inland to be deposited in thin but acceptable rainfall on the slopes and plains.

The same winds send monsoon rains to the countries north of Australia but truly drenching rainfall is rare south of the Queensland border. Australia, the 'Lucky Country', is lucky in its weather as well.

It is not a country of extremes. Its hottest day, January 16th, 1899, could only notch up 53.1°C against the world record of 58°C in Libya. Nor does its lowest reading, −22.2°C, compare with Antarctica's −88.3°C. Against the 400-year drought on the Desierto de Atacama in Chile, Australia's rainfall problems seem minute. But Australia's weather does change – sometimes radically, almost always with little warning.

When the British government cautiously opened up the plains of New South Wales and Victoria in the early nineteenth century the freed convicts and 'squatters' who took up the land found a landscape almost as green and well-watered as the land they had left.

Farming thrived until the end of the century but by 1910 it was evident that the weather was changing. Between 1910 and 1940 rainfall in the marginal lands in the far west of New South Wales dropped 50 millimetres a year. Semi-arid land, previously capable of supporting some crops and livestock, became semi-desert on which barely one cow per acre would find food. Sometimes the border moved as much as 70 miles, wiping out not only single farmers but also whole districts.

Scientists did their best to modify the effects of these dramatic changes. During the 1950s Australia led the world in cloud-seeding (spraying clouds with chemicals to promote precipitation), but the meagre results hardly justified the effort. More ambitious plans to divert the snow water of the alpine rivers resulted in increased inland irrigation and hydro-electric power. The flood-prone rivers of New South Wales' north coast were also partially controlled. At the larger problem of what made Australia's weather so variable scientists threw up their hands.

It seemed obvious that drought – and, inevitably bushfires – followed a cyclical pattern. Insurance statistics show a cycle of about seven years in claims for houses burned by bushfire, with high figures in 1945, 1951, 1958, 1965, 1971 and 1977/8. Dr Robert Vines of the Commonwealth Scientific and Industrial Research Organisation (CSIRO) in Melbourne claimed to find a twenty-year cycle of drought across Australia, South Africa and New Zealand, related to sunspot activity. In September 1980 he published a paper predicting drought in these areas in the mid-1980s.

Yet another cycle exists to change Australia's weather. And it was this which drew most attention in the days before and after Ash Wednesday.

Walter Sullivan, Science Correspondent of the *New York Times*, was one of the first journalists to mention the meteorological phenomenon called 'El Nino' – The Little One or Christ Child – a name coined by the people of the Spanish-speaking South American countries who knew it as a phenomenon which recurred around Christmas every five to seven years. 'Recent sweeping fires in Australia,' he wrote on February 28th, 'heavy snows in the eastern United States, unseasonal rains in the south-western US, catastrophic floods in Ecuador and droughts in Malaysia . . . all are at least partly the result of an oscillation in atmospheric circulation . . .' Sullivan quoted scientists who considered a movement in the area of damp, hot, low-pressure air over Indonesia was responsible for all these disasters. As the low shifted to the east, trade winds were no longer drawn west across the Pacific, bringing moist air to Australia.

'It's like a gigantic see-saw,' commented Eugene Rasmusson, of Washington DC's Climate Analysis Centre.

'The moist air moves eastwards and a higher-pressure area moves in, bringing warmer, drier conditions to Indonesia and Australia. When the pressure of a great mass of air changes, even slightly, in areas near the equator, there are dramatic results around the earth.'

Rasmusson did not exaggerate.

The trade winds faded. The warm surface waters of the Pacific, usually cooled and evaporated by these winds, surged back to the east. At the South American coast the waters divided. Some drove south, engulfing the Humboldt Current, whose cold waters carry food for the fish usually caught by the coastal people of Ecuador and Peru. Not only did their food source disappear but warm water brought torrential rainfall and floods. Birds and marine animals disappeared along with the fish. Warm water moving north diverted winds which usually modified the North American winter. In 1976/7 El Nino gave Alaska its warmest winter in a century but sent icy air down over the United States.

For Australia El Nino means a strong, dry, high-pressure system squatting greedily over the continent. Cold, moist air from the south rebounds and is driven west around the Antarctic continent. Winds from the north and west are diverted above Australia towards the equator.

The Nino of 1982 was more than usually savage and it fell outside its usual Christmas pattern of arrival, striking at its worst in May 1982. Meteorologists speculated that the eruption of the volcano El Chichon in Mexico in April 1982 may have jetted enough dust into the atmosphere to affect the balance.

The worst rains in fifty-eight years engulfed Ecuador and Peru. They also made Queen Elizabeth's visit to California a rain-sodden flop. A total of 16.5 million seabirds disappeared from Christmas Island in the middle of the Pacific. Measurements taken of the water temperature showed that the surface of the Pacific was 11°F above normal.

El Nino had drained away the moist winds Australia

44

needs to survive. Water-filled air hovered on every side but a dome of high pressure kept the continent baking in drought for eleven months. El Nino is seldom predictable in its effects, nor in its schedule. Strong Ninos around Christmas time can be followed by secondary phenomena of almost equal power six months later. Australia greeted the speculation with polite scepticism – all the more dubious because nobody could nominate a possible termination date for the drought, nor suggest how it could be averted.

The Christ Child, like everything else in Australia, was 'out to get you'.

# SEVEN

## *Smoke Jumpers*

AUSTRALIANS EXPECTED fires in the high summer of 1983, though none could have anticipated their dimensions. As the long drought dragged on into the summer of 1982 a few specialist study institutes issued a reasoned warning of possible disaster. One was Melbourne's Chisholm Institute of Technology.

The head of its Fire Research Group, David Packham, pointed out that the forests burned out in 1939 were back to their full growth, offering 'mountains of fuel'. He added, 'the only good news is that in a drought year the growth of grass is light and after it is consumed in early summer the bushfire problem becomes a forest fire problem.'

The Chisholm Institute had earlier in 1982 received a grant of $270,000 from the Commonwealth government to evaluate the aerial fighting of bushfires widely used in America. The Commonwealth's Division of Forest Research also launched 'Project Aquarius' at the same time for similar work on aerial firefighting and flame retardants but most experts were sceptical about the worth of this system.

To Packham, this belated interest was a case of 'too little too late'. 'Australia needs a determination of priorities', he warned, 'just as we determined them after the Hobart fires of 1967, but, this time, we must have some money and resources to solve the problem. In 1969 fire research did not get one more cent!'

The small *Hobbyfarmer* magazine (circulation 22,000)

47

reported Packham's comments in December 1982 as part of an article on the dangers of bushfire. Almost alone among periodicals it pointed to the clear and present danger of a fiery summer. The same issue contained a long article of advice about fires from the farming journalist and lecturer John Randles. This spelled out the obvious precautions to be taken by any family living in the inflammable bush environment:

- Cut firebreaks and fire roads through any heavy forest or undergrowth.
- Check stationary motors to make sure they don't overheat or send out sparks.
- Clean guttering of accumulated leaves.
- Seal any entry points. 'It is the entry of burning material carried with the fire that usually ignites houses in its path,' Randles warned.

The same advice appeared in leaflets handed out by the Bushfire Council and other advisory bodies. It worked well in the case of a conventional bushfire (though relatively few landowners followed even these simple directions, unsupported as they were by the weight of the law). Against fires of the Ash Wednesday proportions, however, it offered no protection at all. In the absence of any systematic national policy of bushfire research and control a thin line of dedicated amateurs offered the only serious protection from fires.

Victoria's Country Fire Authority is typical of the rural volunteer groups. The landowners themselves provide the manpower and help finance the group by fundraising and charity appeals. They buy their own equipment, most of it outdated. In return for leaving their farms and businesses to fight fires anywhere in their district (and this can cover hundreds of square miles), they receive insurance cover for their lives and the equipment they drive.

Of those killed and injured in the Ash Wednesday fires

many were members of the CFA, including eleven who died at Upper Beaconsfield in the Dandenongs when fire overran two antiquated water-tanker trucks.

Australia has no equivalent of the highly skilled and, perhaps most significantly, well-paid fire services of the North American west coast.

Every visitor to the forested areas of America's north-west has seen the bumper sticker FOREST FIRES PREVENT POVERTY on battered four-wheel-drive vehicles. The Forest Service encourages anyone over eighteen to join the 'smoke jumpers' who fight forest blazes. Between five and seven thousand apply every year.

The service trains them and pays them. *Geo* magazine commented in January 1983, 'In a busy fire season, with lots of overtime, it is not uncommon for a crew member to make as much as $10,000.'

American fire services employ all the techniques of high technology to protect lives, property and the environment. Parachute teams are trained to jump in and spot fires for the MAFS (Modular Airborne Firefighting System) fire-quenching C130s. This was the system under evaluation in 'Project Aquarius'.

MAFS was a public relations winner. Film of C130 Hercules transport aircraft bombing fires with 12,000 gallons of water or chemical 'slurry' had wide exposure on Australian TV but the experts were negative. 'The best protection for houses is a proper firebreak,' said one. 'Aerial dropping of slurry is not relevant to saving houses in the Dandenongs or the Blue Mountains because adequate firebreaks built in time would be far more effective. Towns are their own firebreak if people clear up their own backyards.'

Other experts shared his opinion. The Chief Fire Co-ordinator of the New South Wales Bushfire Council, Bill Hurditch, was quoted as claiming that aerial firefighting was only 5 or 10 per cent effective and that American conditions did not apply to Australia.

'It is ridiculous to say that what is good for fire-fighting in the Rocky Mountains should be OK for Australia,' announced another old hand in the same report. He went on, 'Fire control authorities do not want to be importuned by idiotic politicians or members of the public service into attempting mad and unsafe errands when fires are travelling freely in the high, very high, and extreme range of fire danger.'

Only a week before Ash Wednesday the issue hit the news again. Fires burning in Forests Commission territory near the New South Wales–Victorian border ran out of control and destroyed $10 million of New South Wales pine forest.

Local shire president Peter Rogers testily told the *Sydney Morning Herald*: 'Perhaps nothing could have been done, but it has been alleged here that fire-bombing could perhaps have prevented the fire from intruding into New South Wales.'

Victoria's Minister for Lands and Forests, Rod MacKenzie, told the *Herald* that a Hercules, with other smaller planes, did drop fire retardants on the blaze but that the fire ran out of control after two days. In his view, claimed the report, Rogers' comments were 'quite wrong and grossly irresponsible' – surely an aggressive reaction to an idea that was hardly new.

The official position of the Bushfire Council was spelled out on January 19th 1983, a month before Ash Wednesday, in a letter to the *Sydney Morning Herald* from its secretary, T. J. Anderson.

Although well aware of the importance of efficient firefighting equipment, the Bush Fire Council is concerned that calls for more and more modern equipment should not obscure some basic facts about firefighting . . . The only way to reduce the destructive potential of a bush fire . . . is to reduce the amount of fuel available to feed the flames.

Many people disagreed. One New South Wales fire brigade had independently developed a folding reflective canopy of metallised fabric to protect its members as a fire swept over them. The authorities showed no interest in the concept and its use was restricted to a single brigade.

As far back as 1964 CSIRO created a low-priced metal foil blanket which could protect a trapped firefighter for minutes while the fire swept over him. Despite the minimal cost – $75 each – the foil blankets were not adopted as standard protection for Australian brigades. Ironically the American fire service viewed the device with more interest. Adapting and perfecting it they made such protective equipment obligatory for their personnel.

A few members of the public showed equal enterprise. A swimming pool manufacturer, recognising that many fires on the edge of bushfires began from the embers and burning twigs spilling out of the fire 'crown', fed pipes from his pool to a system of tubes on the roof of his house and then offered the equipment commercially. Though he sold a number of such installations, his plans were deprecated by the authorities. They claimed the system might strain meagre water resources during a fire and that a careless owner might forget to turn it off before fleeing from his home.

There was little discussion in official circles of the newest foreign equipment for fighting fires. Most people in the field must have known of new firefighting vehicles like that perfected in Sweden, an all-weather, all-terrain crawler with its own articulated water tank, capable of reaching fires where no roads existed.

At $150,000 per unit the cost was felt to be too high. Significantly the New South Wales Fire Brigade Service, in charge of protecting the vital ecology of Mount Kosciusko National Park, bought such a vehicle in July 1983. By then, however, the damage was done.

Discussion of new firefighting technology centred almost entirely on cost. It was not, argued many, cost-effective to buy sophisticated new equipment or commit to aerial spraying, which meant maintaining a fleet of C130 transports the year round for possible use in a few days of high fire danger. It was, after all, not industry and expensive city real estate under threat from fires, but private homes and small farms. A threat to more politically and economically sensitive property would have produced, one speculates, a very different response.

# EIGHT

## *Blueprint for chaos*

IN OCTOBER 1970 the West Gate Bridge, under construction near the mouth of the Yarra River on which Melbourne is built, collapsed when metal structures failed in a span 40 feet above the river mud.

Thirty-five men died in Australia's worst construction accident. 'Error begat error', noted the report of the resulting Royal Commission, 'and the events which led to the disaster moved with the inevitability of Greek tragedy. This tragic disaster . . . was utterly unnecessary.'

West Gate had at least one positive result. The confusion as police, fire brigade, emergency medical services and helpful private citizens tried to free the men, drew attention to the need for a co-ordinated disaster plan covering every conceivable emergency. Creating such a plan amid the fanatically territorial bureaucracies of Australia was booby-trapped with problems; it was not until 1982, twelve years later, that the Victorian State Disaster Plan was formally unveiled.

South Australia adopted a similar plan in 1980 after that year's catastrophic fires, also on Ash Wednesday, although it required the governor to declare a state of emergency before the disaster plan and its co-ordinator-headed committee took over. In retrospect, considering the effect it was to have on the events of Ash Wednesday 1983, it might have been best if the Victorian scheme, nicknamed 'Displan', had stayed on the shelf.

53

The plan outlined three levels of disaster:

*Stage 1*  Small local incidents.

*Stage 2*  A major disaster, demanding the resources of an entire region.

*Stage 3*  A state-wide or national event, calling on facilities of a number of services.

A 'co-ordinator' was appointed. In February 1983 he was Deputy Police Commissioner Eric Miller. The plan called for no controller: no bureaucracy would risk relinquishing actual power to another individual.

The Victorian plan was in its turn part of a national disaster plan co-ordinated through the Natural Disaster Office in Canberra, which could for instance authorise the use of troops in an emergency.

This plan, untested in practice, would have its first workout on Ash Wednesday and prove a blueprint for chaos. Ironically the National Counter-Disaster College, set up as a training school for officials of the Commonwealth government's National Disaster Organisation, was housed in the converted Golf Club House at Mount Macedon, one of the first towns to be devastated on February 16th.

Ranged against the looming threat of wildfire was a miscellaneous collection of firefighting authorities. The Country Fire Authority, essentially voluntary but enjoying government assistance, technically dealt with any bushfires not in state parks or forests. These were the responsibility of the Forests Commission, which maintained its own firefighting teams and spotter planes. Additionally the Metropolitan Fire Brigades dealt with fires within the Melbourne area.

Inevitably the Country Fire Brigades confronted most bushfires. Their 12,088 brigades and 107,000 members seemed ample for the task, but with those forces distributed across the state it would prove almost impossible to move them quickly between towns when fires ran out of control.

The primitive communications' systems did not help.

South Australian Country Fire Service tankers carry twenty channel radios; those in Victoria – despite representations to the federal communications for more frequencies – had only six or, in a few cases, ten channels. There were no command or emergency frequencies.

An American Fire Chief, Allan West, toured Victoria's high fire risk areas some months before Ash Wednesday. Interviewed later on national TV he was politely dubious about provisions for safety, particularly in the Dandenongs and other heavily forested residential areas.

West pointed out that, while American fire regulations demanded the clearance of most large trees for an area of 400 feet around a house, those in Australia often grew literally outside the windows.

Both the United States and New Zealand require roadside verges to be free of trees and brush, leaving power lines well clear of fire risk. No such regulations applied in Australia, as was to become apparent with tragic results. Embers of the Ash Wednesday fires would still be glowing when furious Victorians made serious and astonishing allegations against the State electricity authorities.

Nor did most towns have an evacuation plan. Displan had advised communities to set up such schemes but many were tardy in doing so. Pakenham in the Dandenongs had its initial meeting on the subject one day before Ash Wednesday.

Cockatoo in the Dandenongs was among the earliest communities to set up a disaster plan. Forty-five people met in the town's play centre on July 16th, 1982 and established a Disaster Planning Workshop. It appointed co-ordinators for first aid, catering and communication and established the centre itself, a new building set in open ground, as an emergency gathering point.

Cockatoo was razed on Ash Wednesday but the plan and the childrens' play centre building were to feature in one of the most remarkable escapes.

The urgent need for planning was dramatically illus-

55

trated all over the fire area on Ash Wednesday. As flames rushed towards their towns people either remained at home far too long, assuming the Fire Service would deal with the blaze or ran on to the roads in panic, sometimes driving or running towards the fire rather than away from it. Their desperate presence strained the already overstretched emergency services.

Who had responsibility for forcing householders to clear their properties of potential bushfire fuel? A ballet of buck-passing ensued as the Country Fire Authority nominated the local councils, the councils their shire sub-sections and shire officers, weary of arguing with irate householders, threw up their hands.

The few council officers who had tried to police their areas met with spirited, even angry resistance. Public meetings were held in the Dandenongs by house owners furious at regulations which, as they saw it, demanded they create in their forest retreats an imitation of the well-mown suburban emptiness they had moved there to escape.

Under the 1964 Country Fire Authority Act the CFA had the power to take over fire control from the councils. It was a power they guarded jealously but hesitated to exercise. Local councils offered them tacit support in this attitude. Most boasted large conservationist lobbies. House owners could not cut down trees without written permission from the council. For the Fire Authority to ask that trees, rather than being protected, be removed in the interests of safety would have sparked a conflict of interest and a test of strength both wished to avoid. The result was a stalemate and the creation of a habitat uniquely fit to burn.

In the angry aftermath of the fires, distraught home owners turned predictably on the councils, councils on the CFA and the CFA on the councils. Perhaps the Fire Authority's argument carried more weight than most, backed up as it was by the grim columns of funeral announcements in the Melbourne newspapers, each preceded by the tiny badge indicating a dead CFA volunteer.

A *Sydney Morning Herald* news story of February 26th led reporter Tony Harrington into a jungle of recriminations. The Dandenongs' Regional Fire Officer for the CFA, Ron Russell, attacked Sherbrooke Shire Council for refusing to allow necessary improvements to, among other things, fire access roads. Equally furious citizens were quoted on the subject of council and shire intransigence.

Mike Hauler, one organiser of a public meeting about the fires, charged Sherbrooke Council with failing to build an essential dam to replenish firefighters' tanker trucks. Another meeting organiser, Nick Van Rossendael, said of the council: 'All it has done is obstruct, constrain and harass the CFA's efforts to make this area safe for residents . . . There is an element in this council which is absolutely irrational.'

The public meeting on February 27th at the Belgrave South football ground charged councils with all this and more. Van Rossendael said: 'The roads have to be widened or they are death traps for both CFA trucks and for residents.'

He also blamed the council for failing to insist on the felling of dangerous trees. A rebuttal from the deputy shire president, Michael Buxton, that, 'in the past three years, 94 per cent of all applications for tree clearing have been approved by the council', met with little applause. It was a time for scapegoats not explanations.

British ecologist David Bellamy, visiting Australia to fight on what was then the primary front in the environmental battle – saving the Franklin River in Tasmania from being flooded by a new dam – backed up the good sense spoken and written for years by local experts like Eric Rolls, Vincent Serventy and David Packham.

Bellamy pointed out that although many local plants were fire-retardant, garden-proud residents had replaced them with inflammable foreign varieties. 'Why weren't there firebreaks around the towns which have been burned into nothingness?' he asked. 'If the areas had been fire-managed and conserved in the proper way, maybe the tragedy wouldn't have happened.'

Such questions don't take into account Australians' passionate territoriality, their reverence for the garden and 'back-yard' as symbols of independence. There were times during the fires when it seemed to some exasperated firefighters and safety officials that the Australian landowner would rather see his property burn than acknowledge he was wrong.

Edward Mundie's book on the Cockatoo fire quotes a revealing exchange between veteran firefighter Harry Innis and a local farmer. With the fire that would destroy the town moving over the hill above the man's farm, Innis asked him to use his tractor's grader blade to cut a firebreak. The man declined – the tractor's battery was flat and he would have to remove a charged one from a fully loaded truck to get it moving. According to Mundie, Innis looked at the fire among the trees and said, 'All right then, Bill. Give us a ring when your house gets alight and we'll come and do something about it.' Later, in the car, he raged, 'Apathy. Godawful, don't-give-a-bugger Australian apathy.'

Most media coverage of the fires stressed the heroism of those who fought and sometimes died to save their homes and those of others, while avoiding outright criticism of poor management, bureaucratic bungling and conflict of interest.

An ABC TV report by Mary Delahunty was almost alone in sounding a critical note. 'On Ash Wednesday', she said, 'Victoria was not prepared for the fires we knew the summer must bring. Bush caressed our houses, and nobody sounded a warning.'

In fact warnings were sounded by scores of people and ignored. The less palatable lesson Ms Delahunty saved for an article commenting on her documentary. 'We haven't accepted that if we are going to live in the bush we have to make compromises.'

# NINE

## *The physics of disaster*

IT WAS, said one writer, 'as if an avenging angel had winged its way through the air, scattering fire brands far and wide, its wake lit up by flaming forests'.

'Darkness fell at midday,' a survivor recalled. 'There was no sound, yet the noise was so great we couldn't hear one another even when shouting. Women ran outside and knelt on the ground and prayed. We thought the end of the world was coming to us.'

Long before the flames there was the smoke – not the grey, shifting clouds a city person would know from bonfires and the occasional grass fire, but smoke more like the thick, impenetrable fumes that gush from a volcano. 'By noon-time,' said one report, 'city-dwellers had left the streets where gusts of smoke and dust had enveloped the town and obscured the light of the sun.'

In the face of bushfire even those intellectually aware of disaster are struck with awe. 'Those of us with the fire falling on us,' said a survivor, 'were unable to equate that moment, when the wad of purple-black matter rolled at us and suddenly burst and blew up in fire around and on us, with any other moment we had heard of or experienced.'

This last survivor was the author Patsy Adam-Smith, speaking of the 1967 Tasmanian fires. The other quotes were from observers not of Ash Wednesday but the 1857, 1898 and 1967 fires. Adam-Smith's article in the *National Times* of February 27th was one of the more reasoned and

59

unemotional of those which appeared in the wake of the 1983 tragedy.

From personal experience she described a disaster which, in all its essentials, seemed more like a divine visitation than a natural phenomenon. For years after a fire people who have lived through it flinch at the crackle of a paper, the smell of a match.

Character changes. 'There was an incredible amount of swearing,' said one survivor. 'More than I've ever heard under any other circumstances. Everyone wanted someone to blame.'

People can appear callously indifferent to the carnage of a fire. Bodies charred to black lumps, limbs burned off, no inch of skin recognisable, earn numb resignation or a belligerent fury. In such relics, little humanity resides.

Animals rate even less tenderness. 'I saw a man kicking his dead sheep,' someone remarked. 'He stood in a paddock and kicked them. Guts were bursting out of charred bellies. He'd just had enough.'

No ordinary disaster creates such emotional havoc. The Australian bushfire, with its apocalyptic assault of heat, smoke, fumes, dust and wind, is unique.

To have some conception of the bushfire's capacity for terror one has to see it not as a simple fire but as a kind of fire storm. Like a storm it has a front in which its primary force is concentrated and its effects are largely atmospheric. Once the proper conditions of dry fuel and hot wind have been created the bushfire burns mostly in the air. One 1983 survivor called it 'a giant blowtorch'. What an apt simile for a force that generates at its front enough heat to fuse sand into glass and soften metal, but which a man can survive if he lies down in the open wearing a reflective blanket.

Most people who die in bushfires are not burned to death. One researcher listed five possible causes of death from fires: poison from toxic gases like carbon monoxide, suffocation from smoke or lack of oxygen, lung failure due to superheated air, burning and the failure of the body to regulate its own internal heat.

'An analysis of these causes,' he wrote 'indicates that only heat regulation failure is a significant cause in bush fires.' Long before fire reaches a victim, he is dead or unconscious from heat stroke or oxygen starvation.

Unlike the smoke of an industrial or domestic fire, that from burning bush is non-toxic, containing none of the poisons generated by burning plastic products. It's 55 per cent solid matter, 25 per cent soot and 20 per cent ash. The levels of carbon monoxide, carbon dioxide and ozone are not much above those in 'clean' air.

A bushfire is also dry. Moisture is quickly vaporised and sucked up into the atmospheric 'crown' above the fire. Dry heat does far less damage to living organisms than air filled with heated steam. Steam heated air at 100°C can scald the throat and threaten the lungs. Dry air at 350°C does mild damage at most. Even at 500°C the risk is far less than that from steam at lower levels.

Authorities agree: you will faint before you choke, choke before you burn. The secret of bushfire survival is to stay in the open and wear clothing that reflects heat. It is seldom safer for a man to run than to stay where he is. Fire can outrun a man, even outpace a car on winding roads. But once the fire has swept over you temperature drops radically. If you have lived through those moments of heat you will probably survive.

The air is hot enough to soften metal road signs which sag in the heat. Cars can survive, however, as the heat is so short-lived that petrol tanks seldom, if ever, explode. 'Stay inside the car,' advises a Bushfire Council pamphlet. 'It's the best shield you have against radiant heat.'

One Dandenongs' woman took this advice on Ash Wednesday and stayed in her car with her child. In the words of one report they were 'roasted'. Another family caught in their car as the Ash Wednesday fire overran Cockatoo lived to describe the experience. Jan and Rob

Fields, their son Mike, a dog and a lamb were trapped in their small van. They drove to an open area away from trees and waited for the fire to strike.

As it engulfed them the air inside began to heat up. Rob doused them with a bottle of beer as they huddled under rugs. When that vaporised he splashed them with the contents of the windscreen-washer reservoir and water from a container providentially left in the van. It sizzled on the metal. But the heat inside the van continued to rise. The metal fittings could not be touched. Jewellery burned the skin. Jan's plastic sandals melted on her feet. In desperation Jan Field tried to get out. The metal door seared her leg and one gust of superheated air forced her to slam the door. It was far worse outside.

When it was safe to get out they found every piece of rubber or plastic on the car had either melted or burned – the door sealing, tail lights and wiring. Tyres had started to liquify and the tubes inside were found later to have rotted from the heat. Despite this the car had given them crucial shelter from the worst of the radiant heat and protected them from the danger of heat collapse.

Some people took these precautions on Ash Wednesday and died nevertheless. But everyone acknowledges that the fires of February 1983 belonged to another realm of disaster entirely. To see why, one has to understand the physics of the bushfire.

Most begin as grass fires, low to the ground and extinguishable with those basic firefighting tools, the water-soaked sack or green gum branch. Once a fire gets into the fuel of a forest environment, however, the dried bark, seed pods and stunted brush generate enough heat to fire the trees themselves.

Gum trees burn with a crazed fury that can seem almost malign. Almost instantly the fire spreads to the canopy of leaves, out of reach of firefighters. Filled with oil, the leaves blaze, releasing clouds of eucalyptus gas. The very air becomes inflammable.

Below the brushfire has ignited tree trunks, whose dry bark burns vigorously. Oil inside the trunk is heated and begins to vaporise. If the temperature remains relatively low, under 300°C, and the wind mild, the trees will merely burn. There is a chance of controlling the fire by bulldozing firebreaks or 'back-burning' – deliberately setting another fire ahead of the main blaze. A slow-moving fire reaching a burned-out area will often peter out for lack of new volatile fuel.

Firefighters always pray that the wind won't change. It is wind that makes the bushfire so virulently destructive – the hot, dry wind of drought country which might have been designed by a maleficent power to fan the fire in a gum forest. If the wind rises so does the heat. It can soar to 700°C. The burning trees no longer burn; superheated by radiant heat the gases inside literally blast the trunk apart and flaming fragments fountain into the air.

With the wind at its back the fire runs amok. As hot air rushes up from the burning forest, new air is drawn in among the trunks below forcing a draught. Flaming debris from trees and brush gushes up into the crown, where it is instantly ignited. The wind sends ahead of the fire a long tendril of flame. Forest half a mile away can be set alight and a town cut off literally in minutes. The ability of the fire to leap roads, even whole towns, contributes to the illusion it gives of being a demonic force, almost an intelligence.

The canopy of smoke and ash rolling in the crown and thrust ahead of the firefront by the wind creates panic in the most experienced bushman. As the smoke dips down in the cooler air before the fire, engulfing a town, it seems that, as so many survivors say, the world has come to an end. Then, racing at them with furious speed, travelling at a velocity nobody who has not seen such a phenomenon can believe, the fire is among them.

Few people understand the speed of a bushfire. With a strong wind behind it the fire can travel at 15 miles an hour.

In hilly country speed doubles for each ten-degree increase in slope and decreases by half for each ten degrees of downhill slope.

A town hit by fire may survive if its inhabitants have taken precautions. Firebreaks keep naked flame from a building and an agile home owner with a knapsack spray or even wet sacks can snuff out flying debris before it has a chance to ignite the house. Larger brick or stone buildings, particularly if set in parkland or amidst paved streets and parking areas, may also survive if the roof can be kept clear of fire.

On two notable occasions during Ash Wednesday, at the childrens' play centre in Cockatoo and at the hotel at Macedon, both of which had hundreds of people sheltering inside, the buildings were saved by small groups of men running buckets to the roof or merely quenching embers with a wet rag.

But the houses most commonly found in fire-prone areas like the Dandenongs stand little chance. Often of wood construction, surrounded by trees and scenically placed on a hillside, they provide ideal fuel for a fire. Their flat roofs, picture windows or open underfloor storage areas welcome the flames.

A fire running wild in a town like Cockatoo invites comparison with Armageddon.

The wind is gale force, strong enough to knock people down and send them tumbling helplessly, the smoke so thick that car headlights can't penetrate it. Striking out of the smoke the front sends gouts of fire, vortices of flaming eucalyptus gas in among the orderly rows of homes. The heat is so intense that the earth itself is burning, the mulch of plant matter smouldering inches below the surface. Soft rocks like sandstone calcine crumble; whole rock surfaces char and break up. Houses don't simply burn in this inferno – they explode as furnace temperatures heat the internal air beyond the ability of walls to withstand the pressure. A corrugated-iron roof, often the least inflammable part of a house, can be blown hundreds of feet in the air and whirled a mile away intact.

The fire that hit South Australia and Victoria on the evening of February 16th could be compared only with the man-made fire storms that immolated Hiroshima after the first atomic bomb or the fire storm that burned Dresden after Allied bombers set it ablaze on the nights of February 13th and 14th, 1945.

The heat in Dresden was over 538°C. Comparable temperatures were generated in the firefront on Ash Wednesday. Bodies were lifted by the draught and flung about in both Dresden and Cockatoo. The heat, the wind, the deafening roar: all feature in reports of the most terrifying fires ever lit by man.

What made Ash Wednesday worse was the fact that nature, not man, was the malefactor. Assisted by freak weather conditions, a four-year drought and an environment rendered as inflammable as a paint factory, the Ash Wednesday fires launched themselves at man with a mindless ferocity he was helpless to control.

# TEN

## *'A tragic question without answer'?*

THE FRIGHTENING significance that Ash Wednesday 1983, would assume in Australian history was hardly apparent until the day was almost over. Even in the devastated areas nobody knew the extent of the disaster until the day after, and even months later experts are still probing the chaos of mismanagement, ill-chance, heroism, futility and sudden death that may yet make Ash Wednesday a metaphor for certain sorts of human disorder.

We knew there were fires. In the summer there were always fires. But nobody took them seriously.

Education about bushfires was minimal. The closest a city-dweller came to acknowledging any personal risk from fires were the signs along country roads warning that the day was one of low, medium or high fire danger. The pointers on these semicircular, segmented signs seemed always to rest in the orange of 'Very High'.

Official bodies like the Bushfire Council and the various Forest Commissions published leaflets, brochures and booklets, but generally it took a visit to the main office to acquire these. The inquest into Ash Wednesday would reveal a total confusion among the various bodies as to where responsibility began and ended.

Not that the public showed any overwhelming interest in protecting itself against fires. Like most disasters, fire broke out because, believing it could never happen to them, people failed to take precautions.

News reports of fires had been so common throughout the summer that by February they barely registered on the public consciousness. Those few experts who comprehended the unique set of circumstances that would soon plunge the continent into chaos were not taken seriously. Nobody realised that the year-long drought, the bushfires, the dust storms and hot, dry winds might all be related in a single environmental process whose results could be predicted and perhaps averted. The thought that Australia's problems were linked to those of other nations around the Pacific basin never entered anyone's mind.

It was to prove a fatal error of judgment.

Ash Wednesday itself, the climax of Australia's experience of El Nino, would be well advanced before anyone realised just how catastrophic the situation was.

The fires began outside Adelaide, the capital of South Australia and, like Melbourne, a low, open city on the British plan, some distance from the sea and ringed by wooded hills in which those able to afford them built holiday homes. Predictably for a city noted for its wealth and culture, Adelaide's suburbs harboured some notable collections and libraries.

Adelaide's newspapers, the morning *Advertiser* and the evening *News* tell the story of inattention, indifference and carelessness. The *Advertiser* of Tuesday, February 15th, the day before the fire, led on political news. The federal election was in full swing, with a newly appointed Labour Opposition leader, the ex-union organiser Bob Hawke, attacking the long-established Liberal administration of grazier Malcolm Fraser – a battle whose outcome was to be crucially affected by the fires.

A story in the middle of page one showed fires raging in the Kultpo pine forest 20 miles south of the city. Others were burning near roads in the heavily wooded area. Almost as much importance, however, was given to the story of 77-year-old Miss Cecile Lanyon's praise of the Country Fire Services for rescuing her forty dogs, which had nearly burned to death when fire threatened her kennels close to the heart of the fires.

The following day, Ash Wednesday itself, fires did not feature on the front page at all. Page three carried two items. One announced that six parks in the Adelaide Hills would be closed that day because of fire danger.

At the foot of the page, police reporter John Whistler filed a report which, in its implications, should have alerted any but the most careless to the explosive combination of circumstances faced by the cities of the Australian southeast. 'The Country Fire Services,' wrote Whistler, 'will consider taking independent legal action against people convicted of lighting fires . . .'

He quoted CFS director L. C. Johns who had said, 'We simply aren't getting the support from the courts. There is a ground swell of criticism by local councils and authorities at courts because they say the penalties for people convicted of lighting fires are not harsh enough. Many are getting away with good behaviour bonds. That's not good enough.' Johns claimed that the maximum penalties, $1000 for lighting a fire on a day of fire danger, and $2000 for a subsequent offence, had never been applied in South Australia.

Whistler then detailed the case of a man whose use of a welder was alleged to have set fire to a large area two days before. The CFS took three hours to control the blaze apparently caused by a few sparks from careless metal work.

Ironic, in view of the accusations made later of failure to provide volunteer bushfire-fighters with protective clothing, was a page ten description of Adelaide Metropolitan firemen dealing with a chemical spill in chemically resistant suits which had cost $2000.

The next day is Ash Wednesday . . .

News of bushfires burning near Adelaide send many of us to our phones to check on friends and relatives. They will be vague about the extent of the fires, perhaps having seen no more than appears in the daily newspapers, printed long before the hour of maximum danger and first disaster, 2 p.m.

That day's Adelaide *News* will give fires only a two-column side-bar – 'Firemen Battle Two Blazes . . . The fires – at McLaren Flat and Clare – were reported to be large . . .' – and headlines will go to an announcement by Bob Hawke of promised tax cuts if his party wins the election the following month.

The afternoon newspapers and those of the next day will tell a very different and tragic story.

Before nightfall fires will be blazing along the southern coast near the South Australian border and between Melbourne and the ocean. Towns in the hills around both Adelaide and Melbourne will be threatened and the volunteer fire services in both states severely over-extended. By nightfall the chaos of Ash Wednesday will rage across southern Australia.

Next day the *Advertiser*'s front page will finally put the news in proper perspective. This time it is the political promises which receive the two-column side-bar, and a bleak headline which dominates most of the rest: '18 Die in State's Day of Disaster'.

Among the related items will be one by reporter Alex Kennedy whose headline raises a despairing sigh from all those environmentalists and fire experts who for so long have been warning of the holocaust to come.

'Why?' it reads. 'A Tragic Question Without Answer.'

# ELEVEN

## *'The sun was going out early'*

A TILTING earth brought southern Australia under the hammering sun at 5.38 a.m. on February 16th, 1983.

El Nino's high-pressure system remained clamped over the continent like a lid, diverting the cool, wet winds which by this time in previous years had offered some relief from the heat of early autumn.

The year before February's weather had been cooler, reaching a maximum of 29.4°C on February 15th. But February 15th, 1983 racked up a temperature in Adelaide of 40.8°C and forecasters estimated a top on Ash Wednesday of 42°.

On the night of the 15th, after Adelaide's Weather Bureau chief, Lyn Mitchell, posted his forecast of the next day's conditions, he prepared a special bench in his office for the Country Fire Services. He had recognised the perfect conditions for a bush fire. 'All it needed was a light.'

On Wednesday morning humidity had plunged to less than 10 per cent – the danger area where the Forests Commission warned that fires could produce flames taller than two-storey houses and move as fast as a man can run.

By early morning hot, dry winds from the parched centre of the continent were blowing south-west across southern Australia. As it had in Melbourne a week before, the wind brought a dust storm over Adelaide, cloaking the city in a gritty shroud.

The first fire was reported at 11.30 a.m. at McLaren Flat, 23 miles south of Adelaide.

The most common reaction was one of disbelief. It was on 20th February, 1980 – also Ash Wednesday – that Adelaide had had its worst bushfires for decades. Although nobody was killed seventy homes were destroyed and enormous damage done to the forest and to crops. The Country Fire Services were still smarting from the long legal and administrative inquiry that followed, part of which showed them as unprepared, underfunded and poorly co-ordinated.

Sure it couldn't happen again?

But the forest had renewed itself in three years. The gums were again full of oil, the ground littered with debris, tinder-dry. Houses had been rebuilt, trees and shrubs tenderly replanted to shade them from the baking summer heat. South Australia was ready, one might even say anxious, to burn.

Twenty-seven minutes after the McLaren Flat fire was reported to the Country Fire Services, the wine country around Clare, 80 miles north, and Tea Tree Gully, northeast in heavily wooded country, was alight.

In the cool, wet, almost European country 300 miles south-east of Adelaide near the border with Victoria fires were spotted at 12.25 p.m. in the huge pine forests around Mount Gambier. But it was eight fires in the Adelaide Hills and the Mount Lofty Ranges that held national attention as the rising wind, now veering to the south-east and south, drove flames up the tree-choked valleys and into the heart of Adelaide's most beautiful satellite towns.

At 1 p.m. the Fire Services declared all eight fires 'out of control'.

In Victoria the day started off no better than it had in Adelaide, and things steadily become worse. A wind blew from the north-east, hot and dry. By 2 p.m., the temperature was 41°C. Forests Commission spotting planes were in the air all morning and they radioed back details of the flash fires all over the state. Before the end of the day ninety-three fires would be logged in Victoria. Ten ran out of control with catastrophic results.

Quick chemical bombing by a C130 Hercules MAFS plane might have doused some of these in their early stages. Only one Hercules was used on Ash Wednesday. Since it flew from a New South Wales base the chance for a speedy response was lost.

Firefighters anxiously watching the weather were told to expect a wind change later in the day, giving relief from the hot north-easterlies that were driving the fires relentlessly south.

Co-ordination of effort, where it existed at all, was casual. Nobody quite believed that a holocaust was already in progress; on the three-level disaster plan this remained a Stage 1 local problem, to be dealt with by either the Forests Commission teams or the Country Fire Authority, depending in whose territory the fire burned.

Would a controller on the American model have saved Victoria from the destruction and loss of life to follow in the evening and night of Wednesday 16th? With hindsight one can see that he might have noted the build-up of an incendiary situation in the environment and moved to avoid the risk of fire by bulldozing firebreaks in vulnerable areas, putting some of the idle MAFS aircraft on alert and co-ordinating the local fire brigades so that at least some could be available in reserve should fire threaten the vulnerable residential area around Melbourne. None of these steps were taken until it was too late. When they were, the result was not co-ordination but chaos.

Just as the disasters of the Australian autumn consisted of not one environmental assault but several, so the Ash Wednesday fires in Victoria were not a single blaze but at least five.

Two major fires burned far to the west of Melbourne, one at Fairhaven and a second on the coast at Framlingham and Warrnambool, where two fires had joined to form a

8-mile front that swept through hundreds of acres of farmland. These are the forgotten fires of Ash Wednesday. Though 181 houses were burned in the first, and eighty-three houses burned and eight people killed in the second, they received less attention than those closer to Melbourne.

Three great fires ringed Victoria's capital on the afternoon of Ash Wednesday.

The first was spotted at 2 p.m. near East Trentham, north-east of the city, near the Wombat State Forest and the towns of Macedon and Mount Macedon.

It is a region of stately homes, set in generous grounds – not a holiday and weekend retreat but more the bastion of Victoria's history. Mansions like Derriweit Heights, Huntly Burn, Matlock, Glen Rannoch and Cameron Lodge stood there like landmarks of Victoria's traditional importance as the centre of law, scholarship and politics. Such houses boasted the great private libraries and art collections of Australia, the cream of the national nineteenth-century heritage as well as evidence of its wealth.

These were no pine-and-shingle weekend cottages, cracker-boxes a match could set alight. Derriweit Heights was up for sale at a reserve of more than a million dollars. The walls were solid brick nearly 2 feet thick, the foundations of bluestone, the roof tile and imported English slate. The amenities included a swimming pool, croquet lawn and 11,000-gallon water storage tank fed from a spring. Surely, reasoned the owners, walls of brick and red sandstone and acres of well-watered, landscaped gardens could withstand anything nature could hurl at them.

They were tragically wrong.

The second fire, in the Dandenong Ranges, started just after 3.30 p.m.

The Dandenongs sprawl in a geological crumple an hour's drive south-east of Melbourne, a complex of wooded valleys webbed with narrow roads. A twisting scenic tourist route climbs into the hills from the satellite

74

town of Ferntree Gully, ending 25 miles from the city in tiny Cockatoo, one of the most beautiful of the mountain villages.

Sawmillers opened up these hills originally and the winding narrow roads recall the bullock tracks on which they were based. In subsequent years a large Italian community moved into the area, growing potatoes and strawberries. Recently farming has given way to residential use as Melburnians took up land and built weekend retreats among the stands of Blue Mountain ash and tall gums.

The main road from Melbourne is mostly two-lane blacktop. It twists through picturesquely named towns like Clematis, Gembrook, Emerald and Avonsleigh. North of the road lies Sherbrooke Forest, filled with the chiming call of the bellbirds, its stands of Blue Mountain ash well-watered and carefully protected against fire. To the south is open country, hilly, in places thickly forested and spotted with tiny hamlets.

Some surfaced roads run through this area but between the winding scenic route and the Princes Highway, 12 miles south, there is little but pasture and the warm, blue-hazed bush. The few towns cling to the highway – Naree Warren, Berwick, Beaconsfield, Officer, Pakenham. Inside the bush area, bracketed north and south by the roads, map names like Harkaway and Narre Warren East often disguise a few shops, a pub and a pole festooned with road signs. The wealthy Melburnians, mostly medical men, who had turned Upper Beaconsfield into a fashionable weekend retreat routinely included a map with their invitations.

The worst Dandenong's fire broke out in mid-afternoon. The thermometer at Meteorological Headquarters had just struck the day's peak of 43°C when, at 3.38 p.m. a fire brigade radioed the Victorian police communications centre D24 with the message, 'We have a large bushfire at Mount Morton Road, Belgrave.'

Belgrave is on the northern, winding road. The actual location of the fire was about 6 miles south of Belgrave at the junction of the area's two major roads – Wellington Road and that between Belgrave and the town of Hallam on the highway, 14 miles south. Locally it is known as the Mount Morton Road.

The fire's curious location, at one of the few points accessible to an outsider in the district, was to become of crucial importance in the days which followed.

In the next thirty minutes, independently, another fire flared in the Dandenongs, 6 miles north-east of the Belgrave fire and just north of the town of Cockatoo. Abruptly, the tiny Cockatoo was thrust into the centre of the stage. It would remain there for the next three months.

The location and circumstances of this fire seemed puzzling also. An official CFA explanation was 'spontaneous combustion'. Seasoned firefighters were not so sure and with reason.

Wednesday's third fire was the most dramatic of all: a blaze by the ocean which tore through a whole district of holiday homes, fire and water meeting in a clash of elements that, to many people who survived the holocaust, was the most frightening and tragic of all.

Far to the south and west of Melbourne, where the same Princes Highway that runs south of the Dandenongs skirts the rocky coast, a series of villages clings to the winding Great Ocean Road. Holiday cottages lose themselves in the oceanside forest. The only sign of civilisation is the occasional general store. People who live in Aireys Inlet, Anglesea, Fairhaven and the areas behind the coast came here, as a newspaper report remarked, 'to retire, or just to escape the bustle of the outside world'. Many own no TV or radio. Not every house has a telephone.

Just after 4 p.m. on Ash Wednesday fire broke out at Dean's Marsh, a small town a little back from the coast near the town of Lorne on the Great Ocean Road.

At about the time the fire started a fox walked out of the bush near Aireys Inlet, north of Lorne, at the beginning of the Great Ocean Road. Disturbed but apparently unafraid it walked around a house, watched by the owner, Chris Condon.

Chris could tell the fox was anxious [reported the *Age*]. It walked all around the house, going up to every door, as if it was looking for something. Chris even wondered later if the creature had been some sort of messenger.

The event was so strange, so untypical for such an animal, which traditionally was so wary of people, that it triggered off what previously had been subconscious in the man. He felt the sensation in his chest.

Psychic tremors began to undo the afternoon, and the sun was going out early.

## TWELVE

*'The sky's red – now it's white!'*

WHILE VICTORIA battled with fires throughout the hot afternoon of Ash Wednesday, South Australia, further down
the road to disaster, was grasping the enormity of its
problem.

One Country Fire Services unit was already out, destroyed when a fire near Hahndorf, deep in the hills, jumped
a road. All CFS personnel in the hills, most of whom had
been on standby at home all day, were called to their base
stations.

Power lines went down and the electricity supply to
Adelaide wavered as emergency systems were tripped.
Journalists trying to update stories of the fire lost copy time
after time as momentary power failures dumped their
computers.

At 2.30 p.m. electrical power to Adelaide failed. Ten
minutes later the South-Eastern Freeway was closed. Fires
were breaking out, apparently at random, throughout the
hills. At 3 p.m. the chairman of the State Disaster Committee and head of the Premier's Department, Max Scriven,
rang Governor Sir Donald Dunstan and advised him a state
of emergency might exist. (He did not actually sign the
papers until five-fifteen that afternoon.)

Communications were breaking down. An inquest into
the fires around Mount Gambier would hear a senior
constable tell the coroner that the telephones went out

79

almost immediately fire hit the area and that he could soon neither receive nor transmit on his car radio.

The coroner asked if distance had been the problem. The policeman replied, 'No. It's mostly due to the standard of our radios.'

The information failure on Ash Wednesday was exacerbated by the fact that few people knew it had taken place.

Misinformation moved in to fill the vacuum. Unwitting radio stations broadcast advice and news bulletins which were either erroneous or out of date. Some reassured parents that all children would be kept in school until after the fire danger. At the same time – three-thirty – sixty children between 12 and 17, students of Glenunga High School, south-east of Adelaide, were climbing into the bus that was to take them home.

Within minutes they were driving between 30-foot walls of flame. Smoke and ashes filled the bus. A 16-year-old girl, Kerry Stirling, calmed them while driver Bryan Blackwell raced through the firefront.

Later a Victorian fireman, his face and hands disfigured with burn scabs, would describe on national television how, in trying to escape in the same way, he ran off the road in a wreck that killed the other man in the truck with him. But the Glenunga bus got through, one of the few stories of Ash Wednesday with a happy ending.

In mid-afternoon Adelaide police urged people with homes in the hills to leave immediately and protect their property – at the same time as country police were putting up barricades to keep drivers out of an area where fires were running out of control. At 3.12 p.m. all main roads into the Adelaide Hills were blocked. Residents made recklessly courageous attempts to reach their families, dodging along back roads to avoid the police barricades. With communications still in chaos, they could only hope the narrow bush tracks were not leading them into the heart of a fire from which others had already tried to escape and failed.

Irene Fenn of Greenhill, a bush suburb a few minutes outside Adelaide, worried about the reports of fire moving into the hills around noon. Her husband Michael had come home when a red alert went out for firefighters and both could see smoke being pushed ahead of the hot winds out of the north.

The school gave its students the afternoon off because of the heat and her two children, Matthew and Rachel, came home for lunch. Irene rang a neighbour closer to the fires. 'There *is* smoke,' she was told, 'but no threat to you.' She decided to leave the children with her husband and keep a doctor's appointment in Adelaide.

She returned to find the road blocked by police. Greenhill had been struck by the firefront that hit the hills and roared up the steep ridge on which the suburb stood. Irene Fenn told a reporter, 'I panicked. All I could think was that I must get home to see if the kids were safe.' She did what many other desperate hills' residents had done – took the back roads. Fire cut her off. She drove to her mother's house to find only her grandmother there. She told Irene her house was burned out and most of Greenhill along with it. But a neighbour had rescued the Fenn children just before the fire struck.

Michael Fenn barely escaped with his life. He had gone to a shed at the rear of his house to get more hose with which to fight the fire when the front broke over him. 'I knew I had somehow to get to the house,' he told a reporter, 'even though it was obviously going to go too. It was a steep climb up but I knew I must hold my breath, because the very air was alight. Before I reached the house I thought I was dead. What a silly way to die, in a bushfire, was all I could think.'

Clothes and hair on fire – the blast of wind blew away his helmet – Fenn got to the house. Limbs from trees which had literally exploded in the heat were crashing into the house. Fenn retreated to his basement until he felt the roof collapse above him, then, wrapping a pair of trousers around his burned hands and face, he clambered into the street. The house opposite, though surrounded by flames, was not burning. Fenn sheltered there with his friend Mike Pietsch, both men darting out to hose smouldering beams or haul away burning branches threatening to ignite Pietsch's house and others in the street.

They survived, Fenn with serious burns on his hands and back.

Few newsmen got past police roadblocks and into the fire area. Adelaide radio journalist Murray Nicoll was an exception. His graphic broadcast to his station 5DN, describing his emotions as he sat watching the last of his house in Greenhill go up in flames, is journalism in the finest tradition – more than can be said for the widely published photograph of Nicoll and his wife Frankie embracing the next morning in the ruin of their home.

Nicoll described the scene in Greenhill with weary determination:

> I am so tired I can't do anything . . . My house is gone. It's exploding still and I just can't bear to look at it.
> And the man across the road from me, he thinks his wife may be trapped in the house, and his house is burned to the ground. Well, if she's in there she's gone. I don't know how many houses are left here; there *were* a hundred.

The manner in which the media chose to report Nicoll's story – portraying him as weary and resigned, a helpless victim of fate – emphasises the tacit agreement by opinion-makers in Australia to view the fires as an Act of God rather than the culmination of a process of confusion and neglect. In fact Nicoll broadcast far more than the fragment reprinted in other newspapers which was syndicated overseas.

After rescuing his family Nicoll found himself trapped with twelve other people behind a farmhouse, being sprayed with water as the fire roared up the slope towards them.

Then it struck.

'The sky is red, then it's white!' shouted Nicoll to the 5AD newsroom over his two-way handset. 'It's going crazy. The fire jumped 100, 150 feet right over the top of Greenhill Road.

'There are something like 120 homes at risk up here. I don't know how many are gone. We are in big trouble.'

Smoke and heat choked off Nicoll's voice. When he could speak again, he said, 'We can hardly breathe. The air is white with heat. There are women crying and there are children here. We are in trouble.'

A newsman on the other end of the line asked if they could send firemen. 'I *am* a bloody fireman,' Nicoll shouted, 'and there are firemen here with me. And there's one Summertown unit gone down the road where the flames came up like an express train. I don't know what happened to it.' The fire brigade survived but five other people died in Greenhill.

Nicoll's experience was shared all over southern Australia that day.

While Greenhill burned fire raced up another tree-covered slope towards the Catholic St Michael's Retreat in Mount Osmond. Brother Joseph watched the fire coming and assembled the six other people at the retreat. But he ran out of time.

The speed was incredible [he said]. I had the car ready to go but when I saw the flames shoot over the ridge I threw the keys on the floor and we all raced down into a small room at the bottom of the house.

It seemed in no time at all the place was full of smoke so we lay down on the floor, put rags over our faces and prayed.

I heard the glass cracking outside, and then the small window we had above us buckled and cracked. Our bodies were heaving and it seemed every bit of mucous was streaming out. And when we tried to blink it felt like shards of glass were fastened inside our eyelids.

Forty-five minutes later Brother Joseph and the others staggered out of the house to find it still blazing, but the firefront itself was hundreds of yards away, destroying yet more old houses which had survived dozens of normal bushfires.

At three-thirty injuries were officially reported. By 5 p.m. confirmation came through of the first deaths, some of them people choked and incinerated in their cars.

In the south-east corner of the state, where fires had been burning most of the day, a man, a woman and four children died on a country dirt road. As the fire approached her farm near Kalangadoo Mary Williams piled her children, the oldest a boy of 7, into her Holden saloon and headed for the farm of Mary Rogers, her nearest neighbour. When they failed to arrive, Mrs Rogers' son, 25-year-old Gavin, went out looking for them. Later firefighters found his burned-out Volvo and, a few yards away, the car containing the bodies of Mrs Williams and her children. He had apparently found them dead, tried to turn back and bogged down in the soft earth at the side of the road. Gavin Rogers' body lay on the road between the two cars.

Fire authorities in both South Australia and Victoria had begun to realise that the fires of Ash Wednesday were unlike the others they'd encountered in the years before. Conventional methods were proving useless. The flames overran not only individuals but whole brigades, destroyed not just single houses but whole towns.

The wind was rising, forcing a draught through the burning trees which turned the air too hot to breathe. Inside such a system the normal rules of combustion broke down. The fired moved like a wall, the trees becoming invisible inside a zone of flame where even the air seemed to be burning. Fire struck a tree, wrapped itself around the trunk, set the canopy of leaves on fire then, driven by a wind of tornado force, streamed forward to the next while the tree juices boiled, vaporised and the whole trunk exploded with a flat, hollow concussion. Leaves, bark and shards of burning wood fountained into the crown of the fire, set instantly alight by temperatures above the melting point of glass.

Birds trying to fly over the fires died in flocks in a breath and tumbled burning out of the sky. Caught in a firefront, car windows exploded and tyres caught fire even though the actual flames were yards away, across a four-lane highway.

Late on Ash Wednesday, CFS director Lloyd Johns was asked to equate the fires with those of three years earlier. 'Five times worse . . .,' he said wearily. 'Six, seven . . .'

As he spoke Melbourne was experiencing the worst night of its life. The fires there were quantified by experts as not six or seven times worse than anything previously known, but thirty times more intense, more destructive and more deadly.

# THIRTEEN

*'The co-ordinator knows nothing about fighting fires'*

WHILE REPORTS of the South Australian tragedy were coming in Victorian disaster authorities struggled to co-ordinate those being flung at them from all over the state.

It was turning into a battle of communications – and communications, as they had in Adelaide, were breaking down. They remained the weakest link in the firefighting line for the rest of the disaster.

Throughout that terrible evening people in the Dandenongs would answer their phones to hear desperate British and European callers ask for news of friends elsewhere in the hills. Unable to reach the emergency numbers they had chosen one with a similar prefix at random.

Home owners in the Dandenongs or among the forests along the Great Ocean Road often knew nothing of the fires until they chanced to turn on the radio or were noticed by a neighbour watering their garden as the front raced down on them.

The Melbourne *Age* told the story of Andie Davis, an English girl who had been in Australia three weeks from a life in Brittany and who was living at Airey's Inlet on the Great Ocean Road in an isolated house surrounded by trees. Late in the afternoon the phone rang. It was a wrong number – but the caller, a local GP, John Eckersley, enquired why she was still there after news of the fire. 'What fire?' she asked.

Eckersley told her that a fire on a 10-mile front was

sweeping towards the town of Lorne, a few miles down the coast. He urged her to get out. When Andie told him she had no transport, he told her to get ready. He'd be there in thirty minutes. By the time Eckersley arrived Andie had had time to reflect. She saw no sign of danger. Even experienced bush-dwellers could not imagine the fire's power then; they believed it would burn itself out, as had others in the past. Even newcomers to the area, when they saw smoke on the horizon, took it for more dust clouds.

She sent Eckersley away.

Andie Davis might have died in Aireys Inlet – one person did and 218 houses were destroyed. But the phone rang twice more – both times friends asking if she was all right. The danger became real. Her near neighbour, Chris Condon, who had seen the fox come prowling up to his house earlier in the day, picked her up. On the way out of the area they were met by a police car. The area was finally being evacuated.

Everywhere in the threatened area people were finding that their vague plans for disaster were laughably inadequate. One of the better prepared towns was Macedon, where the shire council had printed a leaflet giving advice both on the prevention of fires and the evacuation of the town should one occur. Few people took any heed.

The people of Macedon could be forgiven for their scepticism. The area had been struck by fire so often – the most recent occasion being sixteen days previously – that it seemed to many there was not much left to burn. What did remain, particularly the large houses in their wide grounds, appeared impervious to fire. There was an air of justified disbelief, therefore, when fire broke out on the edge of town in a field already hit by fire on a number of previous occasions. It tore through straw-dry pasture towards the hills. Later, serious allegations would be made about the cause of this, in many ways the most disastrous of the Ash Wednesday fires, but for the moment local farmers resignedly slipped into a ritual they now knew well.

In the towns of Macedon and Mount Macedon there was less alarm. Fires had been quenched before. Nobody understood, however, that these fires had the potential to metamorphose into megafires, fire storms with a power to wipe out not only large houses but whole towns, even a city.

The firemen trying to contain the Belgrave fire were the first to know what they faced. Driven by the hot wind it vaulted over Wellington Road and raced into the heavy bush. Brigades on the scene called for reinforcements. The Cockatoo CFA teams had been on standby for days. When the siren rang at three-thirty on Wednesday afternoon, firefighters dropped their work and headed for the fire station. Cockatoo Tanker One was quickly on the road.

Tanker One represented the peak of Australian fire-fighting equipment. With 800 gallons of water and eight pump outlets it could deliver water in spray or jet through a thousand feet of hose. The sides were fireproofed and in an emergency the whole vehicle could be doused from above as the firefighters sheltered behind the heat shields. By comparison with some of the country tankers being called in from outlying districts – many nothing but converted fuel tankers with railed sides for the fire crew – it was highly sophisticated.

On standby at Cockatoo was Tanker Two, a Ford tanker truck paid for by the local brigade itself with $18,000 raised in the district. A high-speed pump made it ideal to back up Tanker One replenishing its tank as it ran dry, but it could also deal with smaller fires independently.

The call to Cockatoo One was from Pakenham, south of the town at the end of the wooded Toomuc Valley. CFA units there were watching the Belgrave blaze head towards them and realising that they lacked the power to deal with a fire on so wide a front.

Firemen trying to slow the Belgrave fire as it bolted through the open grassland and clumps of trees fought gamely but with little hope of success.

The antiquated tools of Australian firefighting – wet

bags, knapsack sprays and water hosed from 500-gallon tanker trucks – were useless. The fuel was too dry, the trees too inflammable, the wind too strong.

Clouds of sparks, whole burning twigs whirled over their heads, starting fires miles behind them, threatening to ring them in flames. Fire roads in this sparsely settled area were meagre. On the already overloaded radio circuits they called for the only tools that would help – bulldozers to crash firebreaks through the bush and rob the fire of fuel.

Everyone dreaded a change of wind. Once a brigade made a stand against a firefront, it was vulnerable. More than once on Ash Wednesday trees burned through and fell on unprotected fire trucks. The fronts were 200 yards long. If the wind shifted, an edge could swing around to trap them in a few minutes.

Meteorologists had promised a change. Everyone prayed for cool winds, perhaps an evening rain shower. Nothing else would affect these fires, except to make them worse.

Many people feel that decisive action might have averted disaster and saved lives. But no such action was taken. Ash Wednesday was regarded until seven-thirty that night as strictly a series of local problems to be dealt with by those on the spot.

It was a time to call up reserves, to summon the many CFA brigades in rural areas. Most were hours away on the winding country roads but some might have reached the Dandenongs or the Great Ocean Road by nightfall when the wind change was expected.

The South Australians brought in troops soon after the fires ran away from them. The same could have been done at 4 p.m. on the afternoon of Ash Wednesday in Victoria. It needed only a call to the Natural Disaster Organisation in Canberra, via the co-ordinator of the disaster plan.

It wasn't done. Nor was the Metropolitan Fire Brigade called out. Not until far too late.

There are explanations.

Idle country brigades were not called because fires were still being fought in some country areas; the authorities feared an outbreak in some distant farming district left without protection while its brigade headed for the city.

The joint fire authorities, Forests and Country Fire Authority, each thought the problem was under control. If either the Chief Fire Officer for the first, Stan Duncan, or Ron Orchard, the CFA's Chief Officer, considered the situation out of hand, neither is on record as having said so. On the disaster plan scale this was still, to them, a Stage 1 – 2 problem, not a perilous Stage 3.

The disaster co-ordinator, Deputy Police Commissioner Miller, was ready to ring Canberra at any time to call out troops or any other resources considered necessary. But, as he acknowledged on national TV, he 'knew nothing about fighting fires'. Nobody from the Forests Commission or the CFA asked for his intervention, so he did not intervene until long after the fires were out of control.

In the middle of Ash Wednesday afternoon, there were still more than five hours before total disaster set in. Fires were burning all over the state and the CFA and Forests Commission brigades were severely overstretched. The Cockatoo tanker teams were only two of scores fighting blazes miles from their own homes – homes they knew were ill-protected against ordinary bushfire, let alone a fire-storm.

If anyone knew what was happening it was the police. They ran the emergency radio network and dealt with crowd control, evacuation and traffic. But, as Deputy Commissioner Miller acknowledged, they knew nothing about fighting fires. And, if the post-mortems on Ash Wednesday may be believed, those who did know – the Forests Commission and the Country Fire Authority – did not always know what the other partner was doing.

# FOURTEEN

## *'That's it – I'm off!'*

ANYONE WHO imagined the Ash Wednesday fires could be handled like any other bushfire abandoned that concept long before nightfall. The vehicle of revelation was the Belgrave fire.

Hot winds gusting due south drove it out of control across Wellington Road, the firefront spreading as its edges found new dry fuel on every side. Its speed terrified even seasoned firefighters.

Within an hour of being reported the fire could be seen from Berwick, 10 miles south, on the Princes Highway. By the evening it would threaten the towns of Naree Warren, Beaconsfield, Berwick, Officer and Pakenham, south and south-east of the outbreak, and Upper Beaconsfield, on its eastern flank.

While the Belgrave fire destroyed a larger area than most others on Ash Wednesday, the fact that it burned in relatively uninhabited forest minimised casualties. It would be among firefighters that this blaze would take its toll.

In this unsignposted wilderness of woodland a single missed turn could send a brigade miles out of its way, often heading into the area of greatest danger. The fire changed its path by the minute. Winds gusting south began to alternate with gusts corkscrewing to the east. The brigades, including Cockatoo's Number One truck, fought to stop the front from spreading over an unmanageable area.

As they did so, fire broke out on the edge of Cockatoo.

Cockatoo's ex-Fire Chief Harry Innis was among the first to sense something was wrong. At around 4 p.m. he walked up the hillside above his own house and looked across the valley. On the far side, near Bailey Road, on the western edge of Cockatoo, a fire was burning. Driven by the winds it was crawling south, 'spotting' ahead as embers and burning twigs carried flames into the desiccated timber.

Innis found a volunteer brigade filling its tank at the fire station and commandeered it to battle the blaze endangering Cockatoo. Tanker Two was already fighting a fire further south down the ridge above the town.

Most of the people in Cockatoo didn't know much more than they could hear on the radio. And the radio said only that a fire burning near Belgrave was also threatening Cockatoo. Not many Cockatoo people took that very seriously. A fire near Belgrave would be driven south not east. Anyway Belgrave was 10 miles away with the Cardinia Reservoir between.

For Cockatoo the afternoon went on normally.

In Officer, south of the Belgrave fire and with Beaconsfield to its west, there was more to worry about. Easterly gusts were carrying smoke across to Upper Beaconsfield.

The greatest risk was to Officer, the highway town south-east of Belgrave. But the front was in check there and at five-thirty, even though the fire's eastward movement was a little more pronounced, the CFA advised the police of 'no apparent danger'.

The fire north of Cockatoo was holding too.

What seemed likely to save Cockatoo and control the Belgrave fire's southward rampage was the geography of the eastern Dandenongs. A heavily forested corridor, the Toomuc Valley, runs between Officer in the south and Cockatoo, 12 miles north. Cockatoo's fire was isolated at the northernmost end of the valley, north of the town. The Belgrave fire had so far stayed west of the valley, whose hills were guiding it further south.

Firefighters knew the risk of fire entering the Toomuc

Valley. The valley itself, with its roads and creek, offered less invitation to a blaze than the forested ridges on either side. A strong wind from either the south or east would send flames ripping along the hilltops with unstoppable speed. As long as the wind remained northerly, however, and the fire at the valley-top near Cockatoo under control, they were safe from that risk.

The hamlet of Upper Beaconsfield, on the eastern rim of the valley, marked a danger point. The country there was a maze of narrow valleys, the houses – particularly two geriatric homes – vulnerable. If the wind continued to push the fire east as well as south, Upper Beaconsfield was at risk and a fire there might spill over into the Toomuc Valley, threatening Cockatoo from the south as well as the north.

Two CFA brigades moved into the area around Upper Beaconsfield: tanker trucks from Narre Warren and Panton Hill. One carried a crew of six, the other five, including a woman.

Panton Hill is a tiny town far to north of the Dandenongs, about 30 miles from Upper Beaconsfield. Its fire crew hardly knew the area. The radio was a roar of overlapping conversations. Smoke filled the sky. They could not have known where they were or which way the fire was moving.

In Melbourne the CFA, Forests Commission and Disaster Organisation may have had a general idea of what was happening, but it is clear that nobody anticipated a major escalation. On the Displan scale, it was still only an emergency at Stage 1, perhaps 2.

Metropolitan firemen – the city service, armed with sophisticated equipment and with 800 trained firefighters among its off-duty men alone – were, they claimed later, rejected if they called D24 offering assistance. Nobody considered it necessary to call for troops or even to put troops on alert.

Police conducted a holding operation, keeping people

out of the threatened areas, maintaining a radio emergency service, helping with evacuation and setting up roadblocks. One such roadblock stood on the Great Ocean Road, barring traffic trying to get to the houses further south, now locked between the inland fire and the sea.

Moving south ahead of the wind the fire would cut the Great Ocean Road at Lorne, the southernmost town on the holiday coast. Aireys Inlet, Anglesea and the other towns on the eastern flank were in no direct danger so far. But the bush behind the town and the homes in it were at risk should the wind change.

To the west the setting sun lit from behind the billowing clouds of smoke. It was impossible to tell what was fire, what smoke, what flaring light. Firefighters raced along the back roads, checking houses, evacuating the last remaining residents.

Some residents, used to fires, didn't want to go. Dame Joan Hammond, the distinguished soprano who had retired to the Great Ocean Road and surrounded herself with the memorabilia of a long career, saw no reason to leave. She was still in her house at 7 p.m. that evening when somebody rang to warn her.

By then, most of the coast had been evacuated. At five-thirty Sue Lucas, who kept a shop in Aireys Inlet, heard the policeman at the roadblock in front of her store say, 'That's it – I'm off.' She locked up and joined the traffic now bumper to bumper along the road to Melbourne.

There was no panic, just the sort of urgency one sees in a rush-hour crowd. Cars inched into the stream from side roads; police calmed the few potential arguments. As an evacuation it showed none of the difficulties the Dandenongs would experience.

Perhaps all this orderly activity lulled emergency authorities into a false sense of security. Even the wave of ash that blew over Melbourne that afternoon from the Macedon fires depressed rather than alarmed the commuter crowds.

Cockatoo road sign.

Aireys Inlet. The fire-storm has reduced even the earth to ash.

Trees near Cockatoo bent by the cyclone winds of Ash Wednesday.

The Great Ocean Road, after the fire.

After Ash Wednesday, Upper Beaconsfield.

Flash floods isolate a house in the Barossa Valley.

Above: South Australian loggers struggle to save a pine forest seared by the Ash Wednesday fires.

Below: The remains of a car swept in the South Australian floods.

Above: Shrewd placement and intelligent architecture saved this house near Melbourne.

Below: Ash Wednesday destroyed the 140-year-old Paechtown cottage of Mrs Sue Arnold.

rvivors huddle on a beach as Ash Wednesday's fires ravage the Victorian
oliday coast.

Cockatoo's Child Care Centre, miraculously intact on the day after Ash Wednesday.

The remains of Derriweit Heights, near Mount Macedon.

After Ash Wednesday, Adelaide Hills.

An RAAF Hercules fitted with MAFS aerial firefighting equipment.

A bushfire rages out of control in Victorian bushland.

he front line of Australian fire-fighting; a volunteer 'wet sack' team in the
)andenongs.

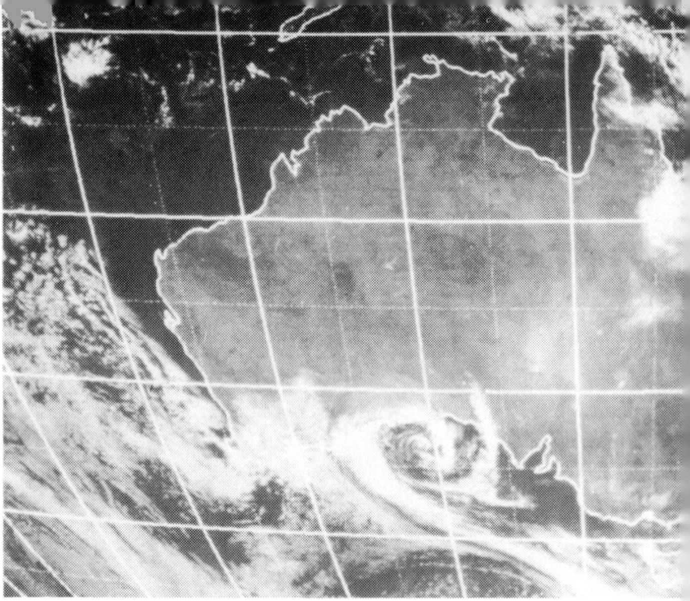

Above: Water-loaded cloud systems
are deflected from the Australian
continent by the high-pressure
system of El Nino.

Below: Cattle, kangaroos and har[e]
marooned by floods at Narran Lak[e]
July, 1983.

Farmers shoot badly burned sheep near Bridgewater in rural South Australia.

83 people squeezed into the tunnel in the Upper Yarra Valley.

Radio journalist Murray Nicoll in the ruins of his Greenhill house.

Above: Peter Gaskell of Hahndorf, in Adelaide Hills, in front of a neighbour's house he tried vainly to [...].

Below: Brother Jeffrey before the ruins of St Michael's Retreat, near Adelaide.

Art collector Kym Bonython before the smoking ruin of 'Eurilla', his 99-year Adelaide Hills home.

The truth was that the fires were already out of control.

Northerly winds were now bearing much more to the east. At the Belgrave fire this forced the flames into a long curve parallel with the Princes Highway. Officer, at the southern entrance to the Toomuc Valley, was threatened. Firefighters found themselves between the highway, with its vulnerable towns, and the fire now burning on a front miles long.

Water was no good against such a fire. Even an area hosed until the fire was quenched and the wood sodden would dry out and catch again within a few minutes. Both air and ground were so hot that they quickly forced out the moisture and apparently dead fires rekindled to cause further havoc. In any event many tankers were now running dry. Bulldozers were called for to cut firebreaks. But those that did reach the roads were stuck around Berwick as homecoming residents, refugees from the fire and emergency vehicles blended in huge traffic jams.

At Macedon and Mount Macedon Forests Commission firemen hadn't been able to keep the fire out of the Wombat State Forest nor save the town of Bullengarook. By sunset it was burning into the area north of Greendale which had been burned out sixteen days before and where fuel was thin. There seemed no risk to the towns or the big houses in the hills. Few cancelled dinner parties. Some people with weekend homes in the area didn't even check on their safety. Radio and TV broadcasts reassured them that the fires were under control.

But already they were breaking through and it was the easterly shift in the wind which was responsible. In the Dandenongs winds from the west began to push the fire east towards Officer and, more dangerously, the hilly edge of the Toomuc Valley around Upper Beaconsfield. On the Great Ocean Road it forced the fire through the narrow distance it had so far maintained from the coast. At six that evening Lorne began to burn.

The fire here anticipated some of the horrors people

would face later that night. Lorne's residents found themselves trapped between the fire and the ocean as their homes started exploding in the flames. Only one refuge existed – the ocean. Wading into the water, they draped blankets over their heads and watched as one house after another fell to the flames.

Construction along the holiday coast, where many houses were mainly those belonging to summer weekenders, relied on cinder block, fibrous plaster and wood – materials that lost their heat quickly in the cool coast evenings. Gypsum board and asbestos-based fibrous plaster, 'fibro', were particularly popular.

Like large areas of glass, plaster board admits the heat of a bushfire with terrifying effectiveness. Houses wreathed in flames heated up so quickly that the air inside exploded before fire had even breached the walls. Sheets of plaster and corrugated iron whirled away in the updrafts as the wind drove the flames on to the bluff above the beach.

Sixty-two houses were destroyed in Lorne but no lives lost. People who tried to fight the flames gave up quickly as they saw their ferocity and speed. This was more than a fire. One Lorne survivor likened it to 'an atomic explosion'. In many of its specifics the firefront *was* like that at the fringes of an atomic blast, though without the shockwave that crushed houses before the searing heat of the fireball set them alight. A closer analogy was as has been mentioned; the fire storms of World War II which ruined German cities like Dresden.

The art of setting a fire storm lay in not burning too much. The incendiaries scattered over a city set fires on every corner and overstrained the firefighters until they gave up and retreated from the area. The fire then began spreading to undamaged houses, fuelling itself until the air became so heated that sand melted into glass and human flesh into fat. Sucking in air from all around the fire created its own draught, forcing oxygen into burning buildings and cracking open those not already in flames. People cowering in basements were cooked, their bodies rendered down until the corpses looked like dolls.

Before the night was out southern Australia would know what it had been like to die in Dresden.

# FIFTEEN

## *Nightfall*

BY EARLY evening even the best co-ordinated and most sweeping emergency action could not have averted the Ash Wednesday disaster. Destruction was merely waiting in the wings for its cue.

With Lorne burning, the Great Ocean Road and the towns of Airey's Inlet, Angelsea and Urquhart's Bluff were regarded by most as unsaveable. No firefighting technique could battle those flames pouring off the hillside and out over the sea.

At Macedon brigades had managed to hold the fire and confine it to an area away from the towns. In the Dandenongs the Belgrave fire was still burning south and east on an ever-widening front and that on the edge of Cockatoo was moving south of the town with new outbreaks reported by the hour.

At five fresh outbreaks appeared both south and north of the fire threatening Cockatoo. Cockatoo's Tanker Two raced to that in the south on Paternoster Road, the main road running down the Toomuc Valley. Before much could be done with the other fire it ran away through the dry grass, reached the wooded slopes of the ridge above the town and linked up with that at Bailey Road and Paternoster Road.

Cockatoo was now menaced to the west by a zone of fire as wide as the town itself. Only the southerly wind kept it from sweeping down the hill into the valley where the houses stood, largely defenceless.

Tanker Number One had been monitoring radio traffic and picked up the news of the new fires. When Cockatoo HQ radioed for them to return they were already on their way. Both tankers began battling the fire on the ridge, determined to hold it away from the town and let the wind carry it south. With luck and without a wind change it would burn away from their homes.

With fires everywhere in the area and dozens of brigades struggling to second-guess blazes that changed their intensity and direction by the minute, communications had reached crisis point.

When Cockatoo One brigade reached the fire they found a brigade from Gembrook fighting on the other side. They had not even known they were there. Monitors advised the Cockatoo firefighters to switch to a less crowded radio channel, but the brigades knew some tankers would miss the announcement of a new wavelength and slip out of touch with the emergency network. Cockatoo refused and the radio became more and more jammed.

Later an operator on the emergency network would recall: 'It was just sheer and utter chaos. We couldn't get through on radio or phone. Communications were virtually blacked out. When we did get through there was no help for us. It's something I don't want to have to go through again.' Later in the evening the police instituted a 'media blackout' on the Dandenongs, resulting in the same confusion and acceptance of misinformation that had bedevilled South Australia earlier in the day.

Amateur radio operators attacked the emergency services later for failing to make use of their informal network which might have supplemented the crumbling police, CFA and Forests Commission radio systems. In retrospect it seems just another opportunity lost – one that might have saved lives.

Commuter traffic was thickening on the roads and police began blockading the fire areas. People returning to their homes in the Dandenongs were stopped by roadblocks. As they had in South Australia, some found back roads and slipped through. Police helicopters cruised the Dandenongs, warning people of danger with sirens and loudhailers. Many householders couldn't hear what the police were shouting. Of those that did, some elected to wait, at least until husbands came home from work.

Those who knew something about fires took elementary precautions – hosing down any exposed wood, such as rafters, filling tanks, baths and guttering with water, dragging garden furniture and other inflammables into the open, raking leaves, placing cars in driveways against the time when they would be needed for a quick getaway. Some went as far as piling their valuables into the car as well.

But years of ignorance, carelessness and lack of education in the effects of fires led in most cases to simple incomprehension. Like drivers convinced they can survive a crash without seat-belts simply by bracing themselves against the dashboard, many people had an unfounded conviction that some simple technique would save them.

For a family of six near Macedon it was the shallow 'dam' on their property. When the fires hit they plunged into the water and waited. A couple in Cockatoo, John Merrick and his fiancée Anne Marie James, took refuge in a creek near the house they were building. Eighty-three people from the Warburton area climbed into the tunnel that carried water from the Upper Yarra Dam. They huddled between two huge pipes as the fire raced towards them. Others fled to their cars, either running from the flames or driving to an open area and hoping for the best.

A few stayed in their houses.

At Fairhaven, near Angelsea on the Great Ocean Road, 64-year-old Colin Barton defied efforts to get him out. 'It's my house and if it burns I'll burn with it,' he yelled at police evacuating the area. Just ouside Cockatoo Yan and Marianna Handli relied on their new brick house and its cellar, the door faced with metal which the salesman assured them would resist fire for an hour. Most people could not consider leaving homes in which they had invested so much. Macedon and the Dandenongs were repositories of Victoria's wealth and tradition. It seemed inconceivable that such a heritage could be destroyed.

At the headquarters of the Disaster Organisation frantic calls between the Forests Commission and the CFA made it clear that the emergency was no longer local. Independently the two authorities made a new assessment of its importance on the Displan scale.

The CFA continued to see it as a Stage 2, as did the police. For the Forests Commission it had already reached Stage 3. At that point such distinctions were probably no more than academic: the time to call in reinforcements was hours past. Later the state Minister for Emergency Services, Race Matthews, would deny vehemently that there had been either confusion or mismanagement on Ash Wednesday night. The assurance, whether correct or not, is irrelevant. The damage was already done.

What triggered a belated realisation of the situation's real dangers was the fire which broke at Warburton around seven-thirty. This is thickly wooded country, some of it state forest. It is also 50 miles east of the city. All available brigades were now committed at the Dandenongs and Great Ocean Road fires. They had run out of manpower.

At 7.30 p.m. the Forests Commission fire officer called his opposite number at the CFA. 'We'll need more men,' he told him. At 8 p.m. the state disaster co-ordinator made his first call of the day for help from the Natural Disaster Centre in Canberra.

He asked for troops and 350 naval ratings were made available. The Metropolitan Fire Brigade's John Perry was scathing about this decision. On national TV Perry claimed that ten spare vehicles and 800 off-duty men could have been supplied just from non-essential equipment and manpower. All they needed was a phone call. Speaking for his fellow firemen he said, 'They can't believe you'd call up 350 naval cadets or ratings who've had no fire training whatsoever, and they be used while we have trained firefighters who are not required.'

The State Secretary of the United Firefighters' Union, Frank Churchill, backed up this charge the day after the fires. 'Up to a thousand men could have been called upon in the last two days. There should have been more use made of metropolitan appliances and equipment. Our members are champing at the bit to help.'

Did petty inter-departmental rivalry lead to the loss of life and property on Ash Wednesday? A remarkable exchange on this matter between ABC TV's investigative reporter Mary Delahunty and Ron Orchard, Chief Fire Officer of the CFA, deserves to be reproduced in full:

DELAHUNTY: So why not call in Metropolitan Fire Brigade Firemen?

ORCHARD: They *were* called in.

DELAHUNTY: But not until late on Wednesday evening, and then only something like seven tankers.

ORCHARD: I don't consider that's a valid comment. Everything we asked for was given to us.

DELAHUNTY: Could you have asked for more?

ORCHARD: You can ask for anything.

DELAHUNTY: Why didn't you?

ORCHARD: Because that was all we required at that time.

The seven tankers didn't get to the firefront until midnight – about the same time as the troops. As to the relative efficiency of the two groups, John Perry remarked later, 'When I was leaving that night the group officer said to me, "Thank you very much for your help, John. You were the only people who went where they were supposed to, did what they were supposed to do, and stayed around."'

In lieu of calls to the Metropolitan Brigade, CFA and

Forests Commission teams from outlying districts were ordered into Melbourne. The nearest uncommitted brigades were in towns like Swan Hill, four hours away by road. When the Macedon fire headquarters called for reinforcements an hour before the town burned they were reportedly told nobody was available.

For the people of the Dandenongs there was no real night on Ash Wednesday. The sky was choked with smoke, stuffed with ash and cinders. Flames lit it from below, creating a dull pink pall that lightened to a livid orange just before fire broke through to engulf a town.

The radios were jammed and in the Dandenongs it was impossible to receive a message clearly. Cries for help had to be sifted from a roar of static and overlapping orders. The police, equally overstretched, could not hold together the uncoordinated emergency services. As the fire to the west ate towards the town they ordered a general evacuation of Cockatoo by private car.

Cockatoo's disaster plan, worked out under Displan guidelines, called for everyone to gather at the pre-school centre, a drum-shaped building of glass with a flat, pebbled roof, surrounded by a grassed open space.

Pre-School Superintendent, Helen Baker looked at the crowds of people, many of them children, already thronging the centre, and ignored the police evacuation order. It was to be a fateful decision.

It would be dramatically apt if the next and most terrifying stage of Ash Wednesday disaster had been heralded by some religious omen, an avatar of doom. But there was none. The northerly winds that had fanned the fires all day merely stopped blowing.

For an instant there was quiet. The flames wavered, though many exhausted firefighters, blind with smoke, choked with ash, noticed nothing.

An instant later, driving out of the south-west, a wave of cold wind struck the Victorian coast. Heat rising from the land drew it across the Great Ocean Road, across the

Dandenongs, across the forests around Mount Macedon and Warburton.

The fires literally exploded.

# SIXTEEN

## *The fire storm*

THE WIND change hit Victoria at about eight-forty on Ash Wednesday night. Roaring out of the south-west with winds gusting up to 40 miles an hour, it swung the fires from a southerly path to one almost due east. The gale-strength Southerly Buster, so familiar to Sydneysiders, that could topple yachts on the harbour or rip washing from a line, effortlessly stripped the crowns of heat, smoke, cinders and ash from over the fires and poured in fresh oxygenated air.

No longer smothered by the waste products of burning, the fires soaked up oxygen and gobbled the new fuel suddenly handed to them. Temperatures rocketed as fire plunged among the trees or bore down on abandoned towns. A CSIRO expert, Vince Dowling, told a meeting at Aireys Inlet that, from the evidence of melted metal, heat there rose to 2000°C.

In burning Victoria fires littered the landscape, but the courage of firefighters had kept large areas within the fire zones intact. Gale-force winds from the change drove these fires together, linking them up across hundreds of yards. Superheated air, loaded with inflammable gases, tore through the forest, moving so fast that people watching a firefront coming had no time even to run, let alone escape. Unburned gas formed balls of violet flame 10 feet across which the wind rolled through the forest at head height as they searched for escape through the blazing canopy above.

At Lorne the wind drove the fire north-west, following

109

the winding Great Ocean Road, while the eastern flank of the main blaze burning in the country behind Lorne swung towards the coast, roaring through the bush to meet the beach like a slamming door. In some places the front reached speeds of 50 miles an hour.

At Belgrave fires eating up the eastern slopes of the Toomuc Valley leaped into the wooded ridges and 'spotted' ahead into the valley itself.

Upper Beaconsfield's 300 residents had been evacuated but the fire engulfed every second house. It moved so quickly that a horse died of heat-stroke and fright before it had time to cross the road.

At Macedon the fire that seemed under control in burned-out country was driven back on itself. Firemen watched helplessly as it swept down on Macedon and Mount Macedon, igniting whole rows of houses and shops, even eating through green foliage towards the garden-set mansions.

Cockatoo's firefighters felt the change and knew the town was lost. Flames jumped the ridge and scorched down into the town. A fire corridor was blown through the heart of Cockatoo and suddenly houses on the far side of the valley were exploding as the heat blasted them.

'It was a fiery wall,' Cockatoo survivor Alan Webb told the *Daily Mirror*. 'You couldn't imagine anything so vicious, so mindless. It roared down the hill like a jumbo jet. We watched houses going up. Poof! Poof! Poof! They didn't catch fire and burn. They just went, all at once.'

Houses simply shimmered in the heat, fumed and dissolved in flame. No fire touched them. The blast of hot air and heat radiated from the blaze sufficed. So savage was the change in temperature and air pressure that windows exploded and flame gushed out through the empty frames.

Metal road signs softened and drooped as if in a furnace. Road surfaces bubbled and caught fire. The earth smoked and changed colour, forming a carpet of hot, soft, pale ash under trees whose trunks, stripped of branches and leaves, seemed clothed in blankets of glowing coal.

Ian Ferguson, in the University of Melbourne's Forestry Department, told the *Mirror*: 'A controllable bushfire can be measured at about 2000 kilowatts of heat energy per metre. This latest fire would be around 60,000 kilowatts per metre. There would be definite similarities with the bomb dropped on Hiroshima.'

Everyone who survived the fires talked about the sound – a roar that deafened, sickened, numbed. At its worst it reached the level of a jet engine howl, but multiplied fifty, a hundred times. Some heard the roar first but could not believe a fire could cause such a sound. A few mistook it for a distant train. They opened the door, looked out, saw nothing but the familiar smoke in the sky and went back inside. Ten minutes later they were racing for their lives among houses that dissolved in flame. There was nothing of fire in this roar, no crackling, popping, not even the concussion of exploding trees. The sound subsumed them all.

John Eckersley, the Aireys Inlet physician, was awed by its power. 'It was just this bloody great force. It wasn't fire by itself. It wasn't just the wind. It was something different to that . . . a monster.'

Men and women caught in this could not fight the fire; they were too busy fighting for their lives.

The tankers were the worst threatened; close to the firefronts, many were trapped when the wind changed. If there was a road near, they reached it and the driver put down his foot while the crew hung on and watched the flames tower high above the trees, reaching up into a crown now filled with debris.

A firefighter trapped in the change near Officer described how his brigade captain took one look at the fire racing at them and abandoned the house they had been trying to save. With only 100 gallons of water left in their tank, they made a stand in the open as the house ignited and a fireball struck a shed 50 feet away. It turned in a moment

111

to flame. The brigade gathered behind the truck and hosed anything within 20 feet that caught fire.

'In an aviary not more than 10 feet from us birds twittered and chirped, petrified,' the captain recalled for reporter Peter Ryan. 'I turned to see our youngest fireman kneeling beside the truck. Obviously ill, he was praying.

'Whoosh! The top of a gas cylinder lifted, shooting flames straight up into a pine, deflowering the foliage in seconds. We stepped forward to quell it. Too hot.'

This brigade survived partly through luck, partly through cool courage. Others were not so lucky.

Even before the fire storm, communications had become so chaotic that the CFA no longer knew with any accuracy how many tanker trucks and brigades were in the area. They officially requested the State Emergency Officer in the Dandenongs, Andrew Helps, to scout the area, counting and identifying tankers.

He found most of them but not, tragically, the Panton Hill and Narre Warren brigades fighting on the slopes of the Toomuc Valley near Upper Beaconsfield. Radio monitors searching through the overloaded frequencies heard a woman calling for help from a trapped tanker. Even if they could have sorted the message out from the rest, they had no reliable fix on their location. In fact the two tankers were isolated on a narrow dirt road running through heavy forest in a steep-sided valley when the firefront swept towards them.

Hoses half unrolled are evidence of the trapped crews' courageous last-ditch attempt to save themselves. But, hemmed in on all sides by dense bush to the very edge of the road, they had no hope of escape. Accelerating up the slope the fire came at them faster than they could run. A few tried, sheltering under ledges. Others stayed in the truck. All eleven died as the oxygen-eating heat and superheated air washed over them. Two bodies were found still in the cab of a tanker, the driver's hands welded to the wheel. Though paint and tyres burned and metal fittings on the exterior melted, the crews might have survived under reflective blankets as used by American smoke jumpers.

Other brigades trapped on the road drove for their lives, the more expert firefighters steering the 16-ton vehicles by instinct on roads rendered unrecognisable by walls of smoke and a torrent of ash and cinders dumped down ahead of the flames. They passed cars blazing by the roadside, tyres and paint alight, windscreens blacked over. Some contained the bodies of fleeing victims, choked to death when the lethal heat washed in through their open windows.

On dangerous curves carcases of cars and trucks lay tumbled like junk below the road. Even if they saw the familiar outline of a tanker among the debris, the trucks roared on. Anyone in that inferno was long since dead.

Burned through, splintered, simply exploded, trees dropped out of the smoke without warning, blocking the road. One could only jam the pedal to the floor and hope to ram the tree aside. Getting out to move such a barrier could be instantly fatal.

Ranald Webster, a survivor of the Nar Nar Goon brigade, interviewed on national TV, displayed a face burned raw from hairline to neck and seared hands healing painfully inside sweating plastic bags. He described driving through the fire storm. 'There were flames the height of the trees. It was like driving through a sheet of flame, right across the road. Not much smoke – just flame.' Unable to see, he missed the road and hit a tree. Struggling out of the back of the car he ran through the flames to safety. His companion, whom he had thought just behind him, burned to death.

No fires disturbed the public consciousness as strongly as those at Aireys Inlet and along the Great Ocean Road. This was Melbourne's holiday coast, familiar to every family. A fire here struck at the very heart of the Australian way of life.

Almost as if the monster of which John Eckersley spoke

had set out to destroy that way of life, the fire launched itself at the coast. In the gale-force wind reality was overturned. Trees bent and splintered or scattered limbs for yard without the fire touching them. A car was forced a hundred yards along a road even though a handbrake locked its wheels. Heavy trucks shook and threatened to overturn as the wind tore at them.

Sheds and garages of corrugated iron lifted intact from concrete platforms and slammed down hundreds of yards away, in one case on top of a fleeing car, which was freed a moment later as the wind sucked it once again into the storm.

Bottles and windows melted in the heat and the sand of the bluff above the beach liquefied to something like glass. Pottery awaiting kiln-firing was baked where it stood in open sheds. The wind ripped flagstones from their beds and moved them a yard away.

Of the houses that burned nothing was left but brick chimneys and these, rather than being blackened by fire, were burned beige and white, the brick surface calcined to ash.

It was not a fire from which objects could be rescued. Steaks would be baked medium to well done in deep freezes, corrugated-iron water tanks boiled dry, then themselves softened so that they sagged crookedly like ill-cooked soufflés. Every machine and implement, from a 10-ton truck to a washing machine, was burned to pale twisted metal.

The best one hoped for was simple survival. The people of Airey's Inlet stood to their necks in the water, blankets over their heads, and watched kangaroos erupt from the 60-foot cliffs over the beach and run crazily down the sand. An instant later flames roared over their heads, streaming out to sea as the fire slammed closed on the coast and began to devour everything that would burn.

# SEVENTEEN

## *'God must love us'*

THERE WAS no lack of courage on Ash Wednesday before the fires, as firefighters risked everything to save the homes of people they did not even know and to rescue those who should have been prepared.

It is impossible to acknowledge fully that spirit because it is not in the nature of these men and women to talk about heroism or sacrifice. Some would return home days later to find nothing there. While they saved the homes of others, their own were burned. Most would say only, 'We were just doing our job.' That it was a job not paid for, taken on voluntarily to protect communities all over the state, not just their own, was seldom mentioned.

There was generosity after the fire too as people uncomplainingly began to pick up the pieces of their lives and those of others. Occasional examples of cheapness, venality and fraud are outweighed a thousand times over by those of almost lunatic open-heartedness.

Before and after the fires these things existed. During the fire storm there was little but stark terror. Everyone realised instantly the fire could not be opposed. After taking one horrified look firefighters abandoned their equipment and ran for their lives. As one remarked, 'Fight that with a knapsack spray? You must be joking.'

Ash Wednesday had gone beyond a disaster. It was now a divine visitation, an attack by nature at her most brutal.

Some wooden houses did not burn. Some made of

115

concrete and stone blazed uncontrollably. It was easy to find evidence that a sprinkler system fitted to the roof saved houses but not all houses fitted with this system resisted the fire, while many that lacked it are still standing.

Ash Wednesday offered convincing evidence that nature has a black sense of humour. A Cockatoo family who loaded all its valuables into a new car and parked it well away from the house would return to find the house intact, unscorched, but the car gutted. For another it was the reverse: the new car was gleaming in the driveway but the house gone.

People survived or died with as little logic.

The family who dived into the dam at Macedon were among those who lived. A brigade crew drove up to the dam to refill its tank and found all six people cowering there. 'They were so scared by what was going on around them,' recalled one of the crew, 'and don't get it wrong – so were we – that they didn't notice the heat was starting to make the dam water boil. We dragged them out. It was as hot as Hades.'

John Merrick and Anne Marie James, the soon-to-be-wed couple who sheltered in a storm water channel near Cockatoo, were found dead in each other's arms.

For the eighty-three Warburton residents who hid in the tunnel between water pipes from the Upper Yarra Dam the decision was luckier. Evacuated from their homes in towns like McMahon's Creek, they took refuge on open land around the dam. As the wind change drove the fire towards them, they climbed down two ladders to huddle 100 feet below ground in the cold, damp tunnel. It was an uncomfortable night; the refugees included three pregnant women, a woman of 70, a baby in arms and fifteen dogs. All of them survived. But Colin Barton who refused to leave his house at Fairhaven was incinerated when it went up.

Yan and Marianna Handli huddled in their basement behind the metal-sheathed door that was supposed to withstand fire for an hour. The whole house above them

116

burned as soon as the fire struck. As the floorboards caught Yan managed to drag open the garage door and they dived out into the heat.

Almost all oxygen had been burned in the air. What remained was close to the ground and they crawled away from the blazing house on their faces. Before heat exhaustion claimed them, they doused themselves from a tub at the bottom of the garden, soaked a woollen car blanket and sheltered under it until the heat was bearable.

In Cockatoo's pre-school centre, 120 people, more than half of them children, had taken refuge when the fire struck. Through the glass walls they could see the fire encircling the open ground on which the building stands.

Just after 9 p.m. Cockatoo's power failed as the poles burned through. In darkness the fire became an all-encompassing vision of hell. People waiting beyond the police roadblocks for news were told by the State Emergency Service that Cockatoo had been totally evacuated and its homes burned to the ground. Residents claim they continued to broadcast this erroneous information long after the service was told that, far from being deserted, Cockatoo still held 700 people trapped.

The town's open spaces saved lives; huddled there around their cars, people could avoid the worst of the fire storm. About forty families sheltered in the gravelled car park of the Cockatoo shopping centre and watched for four hours as the town disintegrated around them.

The pre-school's isolation also protected it from the fire's direct attack but not from the burning coals, branches, cinders and ash being dumped on to the whole of Cockatoo.

Smoke filled the one-roomed building as fires struck the town. Through the glass everyone could see the fireballs rolling through the air just above head height, carroming crazily off buildings or wrapping them in flame. As the terrified people inside huddled with wet towels over their faces, teachers Helen Baker and Iola Tilley calmed the crowd of children, adults, dogs, cats, goats and budgerigars

117

while a few courageous amateur firemen battled to keep fire from the building.

David Adshead, a Cockatoo scaffolder, climbed on to the roof with nothing but a wet rag and slapped out embers as they rained down on him. Later a man played a hose over the pebbled surface, quenching the coals.

A survivor told Jane Fraser of the *Australian*: 'I couldn't even pray. I had this lump of bile in my throat. I wanted to scream but I thought it would make the children so frightened. I knew that any minute we would probably be incinerated and I just hoped that I would die first so I didn't have to listen to the dreadful screams of pain and fear.'

At about three forty-five in the morning a power pole across the road from the centre caught flame. Watching the pole and its crossbar turn to a flaming cross, a child in the darkness called, 'Mummy, look – God must love us.'

The pre-school's survival did seem miraculous. Eighty per cent of Cockatoo's houses in the path of the fire were burning. The fire station itself, crowded with refugees and fire victims, some of them suffering from burns for which the staff could do nothing but douse them with water to ease their pain, was threatened and subsequently caught fire.

Liz Streeter, the wife of a firefighter, manned the radio all night. She wasn't told until later that her own home had burned. She doodled the design for a new one as she worked – it was impossible to take one's own tragedies seriously with so many others all around.

At Macedon, another defensive action was being fought against the fire. At 10 p.m. the fire roared up the Calder Highway from what was left of the little town of Gisborne, leaving 2000 homeless people sheltering in churches, halls and in the centre of the racecourse at devastated Wood End.

Everything inflammable in the path of the fire burned. Along 18 miles of railway between Kyneton and Gisborne almost every sleeper was blazing. One awed resident said of the fire as it raged towards Macedon and Mount Macedon: 'It must have been going at 50 miles an hour, 50 feet high.'

'Two people I knew were killed,' he told the *Sydney Morning Herald*. 'An elderly lady – the nicest old lady in town – ran out of her house to try to get to the hotel, and was burned to death. A man in a car missed a corner and went into the trees.' The final death toll in the area was eight.

The fire seemed invincible. Whole streets were in flames. The CFA headquarters caught. Two hundred houses would be destroyed in Macedon, 150 in Mount Macedon. Three hundred of Macedon's 1000 residents retreated to the hotel, a modern brick building less likely to burn.

Twelve firefighters took up station on the main street and, while the buildings opposite blazed and cars exploded like bombs, fought eyeball to eyeball with the fire. They held the line in front but a torrent of blazing coals roared over their heads to cascade on to the roof and the hotel outbuildings. The eaves caught. They were snuffed out and wet towels stuffed under the doors to keep out sparks.

The flat roof was vulnerable. Publican Brian Nish leaned a 20-foot ladder against the back wall and became a one-man brigade, filling two plastic buckets from a swimming pool in the backyard, climbing on to the roof and dousing the hundreds of red-hot coals rattling down from the smoke-filled sky. He emptied half the pool through the night – 1500 gallons of water. The hotel was saved and with it the hundreds of people inside.

Mount Macedon, the small town 3 miles higher in the hills, was less lucky. Half burned-out sixteen days before, it was still dazed from that assault. Ash Wednesday caught it unprepared.

A prominent QC, Clifford Pannam, owned one of the great houses of the Macedon area, Huntly Burn. A thirty room mansion, it contained the law library of a lifetime as well as antiques, a major collection of art, and was set in one of the most famous gardens in the area. It had survived fires before. Pannam had no reason to believe it would not do so again.

Like many people, Pannam first heard the fire about seven-thirty – a roar which he took for a train from Woodend. By 10 p.m. the road out of the area was thick with cars fleeing the fire. Mount Macedon had no firemen to protect her; even if they could have disengaged from fires blazing all over the area most would have been too exhausted to defend their town. For two weeks they had worked day and night to snuff out the remnants of the last blaze.

At 10.30 p.m. the fire roared towards its companion town with Macedon a smoking ruin behind it. 'It came like a ball of flame and went straight up the mountain, said Don Tarrant, licensee of the Oriental Hotel, 'and it finished the town off.'

Within minutes the state school, the Uniting Church, the Country Women's Association, a milk bar, an estate agent's office, the Swan and Perch restaurant and the post office were gone. Only the hotel and the garage next to it survived.

Huntly Burn was razed. Dr Pannam and his family fled as a firefront came exploding out of the south-west before the howling gale-force wind. Again the firefront outstripped its ability to consume the gases it created; what struck Mount Macedon was, as all later agreed, a fireball rolling before the gale. Pannam and his family got out half a mile ahead of it.

They returned to find Huntly Burn, in the *Age*'s description, 'a vast ruin, white with ash [that] bled grey smoke into the sky. Nine high chimneys stood in the twisted corrugated iron and red-hot embers . . .'

Evidence of the Mount Macedon's fire's awesome but random force in its uphill rampage was clear in the ruined grounds of Huntly Burn. A large walnut tree was uprooted, but a plum tree nearby survived, fruit still hanging ripe on its boughs. Next door a variegated elm had exploded, scattering branches for yards around, while a more delicate weeping elm remained untouched.

Derriweit Heights, the mansion at Mount Macedon, burned as well, its bluestone foundations supporting now only a two-storey shell. That it could so lightly blast a house of such solidity totally destroying the history it represented showed again to Australians the casual power of the force of nature against which they had set themselves.

It was drought and fire, not man, that were the dominators in this environment. As Thursday dawned and the newspapers filled with reports of what seemed an apocalyptic assault on southern Australia, people began to count the cost of the environmental Vietnam in which they had involved themselves.

# EIGHTEEN

## *The cost*

THE FIRES of Ash Wednesday burned through the night, largely unopposed. Numbed by the fire storm, firefighters did their best to help the injured and protect those who had found shelter. Ambulances got through to Cockatoo during the night and took out victims, though the roads were still clogged with cars and the verges with exhausted families sleeping under what cover they could find.

As the first dead were brought in, Cockatoo's CFA post set up a temporary mortuary. But many bodies would not be recovered until the morning when rescue teams set out to check burned-out cars littering the district.

The cold, moist south-westerly wind blew steadily all night, dragging down the temperature until those huddled on racecourses and parking lots shivered in the chill. In a few places brief rain showers soaked the weary survivors. Ironically a wind like this, full of moisture, would have been welcomed in the months of drought before February. But Australia remained sealed under El Nino's dome of high pressure: Ash Wednesday's gale was a single exception so disastrous it might have been planned.

Perhaps some meteorologist noted the direction and moistness of the wind and wondered if it was just a freak of nature or perhaps something more, a harbinger of the change everyone prayed for, an end to the year-long drought. However, the fires expunged all other thoughts from the minds of Australians in that appalling February. It would be weeks before the experts saw the wind of Ash Wednesday as the messenger of a new and incredible manifestation of El Nino's power.

After the first explosion at 10 p.m. and the rampage that followed, most of the fires had been driven into new territory, much of it luckily unpopulated. With their first fury exhausted the fires burned on but with less of the explosive intensity that accompanied the wind change.

From their roadblocks police could see the towns on the Great Ocean Road burning in the night. Great houses, their structure eaten out by the fire, collapsed and a wave of sparks washed up into the darkness. Much of Australia's priceless literary and artistic heritage went up with them.

Melbourne's hospital burns units were filling up with people invalided out of the fire zone. Some would spend months enduring the agony of healing and skin grafts, mostly on face and hands – areas a simple reflective suit or blanket would have protected. Another lesson for the ill-prepared emergency services.

By 3 a.m. the Dandenongs' roads were almost clear. A stream of emergency and police vehicles moved into the area, the precursors of a flood to come. In the homes of friends, in makeshift shelters or in cars, evacuees tried to sleep and wondered what had become of their pets, their homes, their friends, their families, their towns. At dawn they began trickling back.

Towns don't detain a fire for long. The blaze that destroyed Cockatoo was burning beyond the next ridge and that which had razed Macedon and Mount Macedon moving through open country miles beyond the towns. So savage had been the fire on the Great Ocean Road that literally nothing remained to burn. Hot ash carpeted the entire coast, studded with the smoking remains of houses, now almost indistinguishable from the grey earth.

In South Australia as well night dulled the fires and drove them away from ruined towns into heavy bushland, where they continued to burn, as they would in rural areas, for days.

At dawn, in the blackened fire zones around Adelaide and Melbourne brigades cruised, hosing down still-smouldering stumps, telephone poles and houses, checking burned-out cars for the grisly remains of the dead.

Everywhere there was a monochrome drabness and chilling quiet. 'It's another world;' said one report, 'where the only colours are black and grey, and the silence can become unbearable.' After an ordinary bushfire animals and birds that have run before it drift back into the burned-out areas before the ashes are cool. But people on the periphery of this fire described entire flocks of birds falling in flames out of the sky and bush animals flinging themselves across freeways in their terror. Nothing stirred on the Thursday after Ash Wednesday. Driving through the blackened remains of the forests one was conscious only of an eerie silence.

In the burned-out areas west of Melbourne landowners were out at dawn struggling to asses their losses. Hundred of miles of fencing had been burned. As for the stock, what remained was often so badly injured that it had to be destroyed. Farmers started cruising their properties in their small utility trucks with loaded rifles, looking for animals in torment.

Simone Kelly of the *Naracoorte Herald* gave one of the most graphic descriptions of the ruin around Lucindale in South Australia. She saw utility trucks loaded with dead sheep:

The black masses had blood running from bullet holes. The carcases were thrown into pits.

Stiff black sheep which you think must be dead are still standing where they were struck, refusing to lay down and give in. Flocks are dead in paddocks and yards.

Guns fire non-stop as more stock fall. Men throw the animals onto utes, trying to ignore the sight and stench.

Firefighters look at you through red slits. Is it from smoke, dust, exhaustion or tears? . . . Many have not slept for 36 hours. When will they sleep? When will it end?

Farmers who lived with fire all their lives fell into a familiar routine, described by the survivor of an earlier rural fire:

They dig a great big hole in the district. Three or four properties would all bury their stock in the same pit. And they'd just shovel it in. It was like those horrible pictures you see of the Nazi death camps. They just push dead things in on top of other dead things.

The animals are in all sorts of bizarre shapes – feet twisted, mouths wrenched open with teeth exposed in macabre grins. It's just vile. I think that's worse than the fire itself. At least you feel that's out of control. There's nothing you can do about it.

Around Melbourne and Adelaide the desolation left people shocked. Few survivors, even those already exposed to bushfires in the past, could believe the extent of the damage.

At Aireys Inlet the earth was hot to a depth of 6 inches – too hot to touch. It remained that way for three days. 'I didn't think earth would retain heat for that long,' said Dr John Eckersley, 'but it did.'

Thursday was another hot, airless day, a day of maximum fire danger. Exhausted brigades drove themselves to patrol the fire areas, digging out smouldering stumps, dousing the roadsides until the last risk of a new outbreak was gone.

The term 'to lose everything' assumed a terrifying reality as Melbourne and Adelaide assessed the damage. These fires left no souvenirs, no reminders of the old life. Everything was incinerated as if in a furnace.

Those burned out picked through the cooling remains of

their houses with incredulity. Between the flaking bricks of the foundations and the layer of crumpled corrugated iron that lay on top of them there was nothing.

Here and there notes and signs signalled a pathetic urge towards community. Stuck to a surviving wall in Aireys Inlet was a paper lettered in red: 'Big Black Wolf Dog found bottom of Boundary Road near beach. Please feed and water.' Underneath, someone had written, 'Dog shot 18.2.83.' In Cockatoo a note pinned to a tree said simply, 'Bob, ring Joe.' 'He won't be gettin' no call,' a CFA man remarked. 'Bob's the bloke they found under a tree. Now he's under a blanket.'

Exhaustion and shock did more damage to many people than did the fires. Men and women wandered dazed in the wreckage; their descriptions of what happened to them on Ash Wednesday would falter, then stop entirely as the horror came back to them.

Australia's fragile heritage of imported art and culture suffered an appalling blow. Some local works may be replaced but rare books burned in Adelaide's Sacred Heart Mission and the French and German antique furniture destroyed at Mount Macedon are just a fraction of the foreign objects no local collector or institute could now afford to acquire.

Dame Joan Hammond returned to the Aireys Inlet home she'd fled fifteen minutes before the fire to find wooden pillars and rolled steel joists framing a picture of desolation. Her arched picture window and balcony still looked out on the blue Pacific, but nothing else remained.

Five thousand books, a priceless library of opera recordings, annotated scores and souvenirs of her long career, including her gold record for the 1941 recording of 'Oh My Beloved Father', were all destroyed.

Jazz entrepreneur Kym Bonython lost not only his house near Adelaide but his collection of 5000 jazz recordings, films and books, as well as paintings, sculptures, tapestries and ceramics by Sidney Nolan. The collection has an estimated value of $750,000 but in real terms its loss cannot be quantified.

Equally impossible to place a value on were the papers of Dr John Playford, the research material for books on four generations of Playfords prominent in Australian politics. All were burned when Drysdale, his 124-year-old house, was burned at Norton Summit, outside Adelaide.

The Dandenongs and the Great Ocean Road had attracted many of Australia's artists and collectors. They lost everything as studios and holiday homes were razed. No count will ever be made of the works of art, the books and manuscripts, scores and sketches consumed by Ash Wednesday's fires.

The *Bulletin* was right to comment; 'After the South Australian and Victorian fires, the local people will rebuild. But they will not be able to recapture the essence of much of what has gone.'

Home owners drifting back into Cockatoo and Macedon and the Adelaide Hills had no such high issues on their mind. What faced them was the task of getting back on their feet again and rebuilding in the ruins of their homes.

The most common reaction to the ruin was a somewhat unhinged hilarity. Publicans retrieved unexploded kegs from their cellars and broached them for passers-by. Shops that had stock opened and stayed open, extending credit where it was needed.

A Telecom engineer arrived at Aireys Inlet on Thursday morning and installed a phone in the only open shop. Government bodies, private companies and individuals came together in an outpouring of generosity typical of Australia at its best that was to continue as Australia and the world grasped the full extent of the Ash Wednesday tragedy.

Not all public response was benign. Brigades were still mopping up at Aireys Inlet when people drove down from Melbourne to ask if there was 'any cheap land' for sale.

A baker's van drove into Cockatoo on Thursday morning and its driver attempted to extract twice the normal price for his bread, using the fires as his excuse. Furious shopkeepers rang his boss – to find the man had been sent out with orders to give the bread away.

The police insisted on a media blackout of the area, which was rigidly enforced – but they were powerless to stop the looting that started almost before the fires were out. Men were arrested at Port Campbell, south of Melbourne, at Princetown and at Emerald in the Dandenongs. Many others were never caught.

One resident whose house had not been damaged said, 'The only reason I've come back to my place is because of the looters. It's bad enough that people should lose everything without these types being around. I hope the police catch them before the residents do. I would hate to think what would happen to them.'

The greed was, however, far outweighed by a remarkable generosity. Tradesmen arrived in Cockatoo ready to work tirelessly and without pay to reconnect water and electricity. The tent city that sprang up offered a bed for people who had not slept for two days, food for those who had not eaten and clothing, bedding, furniture, kitchen utensils and toys for those whose houses were among the 80 per cent destroyed in the fire's path.

If the generosity was typical so, unfortunately, was the search for scapegoats. For that role there was no shortage of candidates.

# NINETEEN

## *Journal of a plague year*

THROUGHOUT THE fires I had stayed at home, like most of the country, and followed the news on TV and in the papers. The sudden cessation of fire stories a few days after Ash Wednesday puzzled and alarmed me. Was that it? The end of the event?

There was no sign yet of an end to the drought that had brought on the fires; less reference still was made to the climatic conditions, the effect of El Nino which, if all the literature was to be believed, would continue to have as baleful an influence over Australia as it had over the countries of South America now, ironically, suffering from freak rain storms and floods.

What would happen when the influence of El Nino did end? It was a question one imagined the experts should be asking. But in the flood of comment on the federal election and the lavish fundraising to help the victims of Ash Wednesday few people seemed interested in taking a reasoned view of our ecological crisis.

Comment on the fires in the press was, for the most part, emotional and melodramatic. It was an event to be lamented not analysed. Soon enough scapegoats would be sought and the inevitable inquests and public inquiries elicit the expert testimony on which future policy might be founded.

If Ash Wednesday and the events leading up to it were to

be understood, they should be most explicable at their source.

I decided to drive to Melbourne to the scene of the fires and on the way to learn something of the disaster Australia seemed so sure was over, and was so anxious to forget.

Friends in the Dandenongs outside Melbourne – the site of the worst fires – told me over the phone something of what to expect. None of what they said seemed to have appeared in the press. As usual the stories for public consumption merely recycled conventional attitudes and confirmed existing prejudices.

I left Sydney at dawn for Melbourne.

In a year of drought no sight on the Australian roadside is more commonplace than a dead kangaroo. The humped, grey-brown corpses were huddled along every mile of the two-lane blacktop Hume Highway that spanned the 600 miles between Sydney and Melbourne.

About noon, somewhere out on the plains, I pulled in under a roadside gum a few yards from yet another body. I knew how the kangaroos died. Drawn at night to the moisture condensed on the asphalt, they hardly noticed the speeding cars until the high-beams spotlighted them. Their eyes met the lights in one dazed, startled stare, an instant before the fenders flung them, crushed, on to the dusty verge of the road.

Drivers could have avoided them if they had wanted to but in Australia no native beast arouses less respect than the kangaroo. Koalas and wombats, bumbling, Disney-esque and cute, earn a grudging affection; the ambling kangaroo, awkward and stupid both, as well as a competitor with the imported cattle and sheep for the precious grass, is to most rural Austalians fit only for extermination. Farmers execute them literally by the million with government agreement. It's not unknown for country sportsmen, weary of the ease with which the kangaroos can be shot in the spotlight, to go hunting them with axes, driving along side the fleeing animals and decapitating them on the run.

I walked back to the body.

Killed early that morning, it had lain in the rising heat of the day, decomposition keeping pace with the soaring air temperature. By 7 a.m. it would have been in the low 20s centigrade. Now, at midday, it was more like 37°C in the motionless air. Blue-winged blowflies droned over the mashed, bloody fur of the crushed forequarters. One promenaded along the lips peeled back from the sharp, nibbling teeth.

In death it looked as ungainly as in life. No Australian animal moves with grace and the kangaroo even less than most. The contradictory demands of its environment dictate, as with so many other animals and plants on this island, a range of survival skills. If a camel is a horse designed by a committee, the kangaroo is a deer designed for a desert.

Feeding, the kangaroo lumbers, hopping on the lower lengths of its backward-bent legs. The heavy tail drags. Held delicately over its sparrow chest, the vestigial forepaws are webbed with veins the blood in which is cooled by the saliva the animal dribbles constantly over the arms and legs.

At rest and when feeding the kangaroo earns only derision or hatred from farmers who see their grass and crops despoiled by the voracious herds. But a scare and the kangaroo is off in the only mode that gives it grace; the bounding leap of full flight, body thrust forward, arms tucked in, tail elevated to counterbalance the weight over muscled haunches that send it tirelessly out of reach of any but the most persistent predator.

The aboriginals traditionally hunted kangaroos with fire, using the blaze to drive them towards the waiting spears. No other predator could catch them until the white man arrived with the fast light truck or faster bullet of a .303 rifle. Now the Australian federal government was about to authorise the extermination of three million kangaroos as being a danger to rural industry.

I looked around at the country in which I had halted. The air was dry enough to make one aware of one's eyeballs and blink faster to lubricate them. Away from my air-conditioned car my mouth had gone gummy dry. The sky was cloudless, almost white. No appreciable rain had fallen here for more than a year.

Once savannah and forest, this land had been cleared a century before to accommodate wheat and sheep, both painstakingly bred from hot-weather strains to survive in a land where rain fell at the rate of only a few inches a year.

A field just beyond the road was given over to wheat. It stood yellow-gold in the sun, a fine crop, but with no stalk taller than my knee. I knew that the national harvest would be half that of previous years. Nationwide the sale of farm machinery was down 27 per cent. In this area it had ceased entirely.

On the other side of the highway a heavy wire fence marked a sheep pasture. The field sloped away to a dip, where a clay-floored pocket had been scooped from the slope to catch run-off rain water. The 'dam' was nearly empty. The surface of what water remained caught the light with a shimmer of green iridescence like a blowfly's wing.

I looked for sheep, searching the shadow of a grove of gums. Only one animal was visible, an unshorn lamb, its wool dried and baked the dusty grey colour of a rotten twig. Too weak to step over the gum roots straggling across the rock in their search for earth and moisture, it stood help-less, as much enclosed as if it had been fenced.

By European standards the gums were ugly, their trunks deformed by bulges and knobs. Fresh white bark alternated with areas of charring. I could see the same mixture of new growth and fire scarring on other trees along the road and scattered through the pasture. They had survived much. Drought and fire left them mutilated but alive. The sheep,

134

on the other hand, would live no longer than a day or two. And the kangaroo was already dead.

Needing food less than a change of view, I detoured from the main road into Canberra, the national capital. The road in skirts Lake George, in better times a popular beauty spot. Picnic grounds shaded by trees still look out on a flat, dry plain that was once the floor of a lake. A rusted car body defines the high-water mark. Yachts raced here. Men drowned where now a few sheep foraged for the last green grass in the district.

Canberra's public buildings – the Palladian National Library, the futuristic cement works of the National Art Gallery – recline by a vast lake, marble façades fanned by the plume of a fountain that jets hundreds of feet from the water. I ate and left, uncomfortable in such luxury when a few miles away the sheep tottered and the hunters cruised with spotlights after what remained of the kangaroo flocks.

Fifty miles further on I came to a hand-painted sign leaning against a roadside post. The silhouette of a sheep and the enigmatic word 'Ahead' gave all the information its painter considered necessary – and all that a countryman needs.

Around the bend hundreds of sheep blocked the road, filling the air with soft gold dust. A drover on a pony eased them along, lazily south. Desperate for forage, he'd taken his mob (or that of an employer) on the 'long paddock'.

Further south again the radio announced that graziers were furious at the government of New South Wales refusing to let them pasture stock on the protected alpine environment of the Mount Kosciusko National Park, only now recovering from the merciless over-grazing of a few years before. Farmers who could not afford the 'long paddock' were hand-feeding, selling off stock, even killing them. The climate was winning the battle everywhere I looked.

In the heart of the drought area I topped a rise to look out over a plain as sun-struck and desolate as something from

135

de Chirico. A difference in the sky hardly caught my eye at first, so dusty white had it been since the sun came up that morning. But now I saw clouds, too high for rain. Their undersides were stained yellow-gold, like the opalescence in antique glass. Between cloud and ground hung a mile-long screen of grey, too thick and unswerving to be rain.

The country was on fire.

## TWENTY

## *Cockatoo*

MY FRIENDS live on the edge of Fern Tree Gully, technically part of the Dandenongs but on the city side, half an hour from Melbourne. It is possible to look back from the hill out of which their wooden house is cantilevered and see the city in the distance.

Beyond this elegant little half town, half suburb the mountain road rises through Emerald and Belgrave Heights until it reaches Cockatoo. I hoped to get through to Cockatoo the next day. The roads, blocked by police barricades for some days, were open again.

Meanwhile it was good, after a drive through that desolate, desiccated landscape, to relax with the sort of luxuries even people of relatively modest resources in Australia regard as staples. It was the chance to enjoy this way of life in semi-rural surroundings that drew so many people to the Dandenongs, to the Adelaide Hills, to the Blue Mountains outside Sydney – three of the worst bushfire areas in Australia.

I asked my friends about their experience of Ash Wednesday night.

They were watching TV in the large sitting-room at one end of the house. 'Irene got up around nine and went through towards the bedrooms. She came out and said, 'The house is full of smoke!'' We looked out the window, and of course the whole city was just blanketed.'

What they saw was smoke blown off the fires by the wind

137

change. Thick with unburned hydrocarbons, the debris of a million incinerated trees, the cloud drifted across the city its underside tinted crimson from the flames then roaring into the fire storm that was to destroy Cockatoo, Beaconsfield and Macedon and Mount Macedon to the north-east.

Someone driving home earlier that afternoon and seeing the smoke moving slowly across the evening sky would describe it as like a vast eyelid. At nine the sky had become a giant bloodshot orb and the smell of burning eucalyptus filled the city.

Next day I drove to Cockatoo.

The brochure handed out by the tourist authority paints a picture of the Dandenongs redolent of elegant middle-class ease. Under 'What's On', it suggests: 'Watch the craftsmen at work – visit the Potters' Corner.' There's a rhododendron festival, an edible crafts festival, art galleries, souvenir shops and restaurants.

Much of the scenic route lies within Sherbrooke Forest. The brochures stress that this is lyrebird country; their silhouettes decorate every road sign and you are warned not to be surprised by the sound of a branch snapping or a bird call you don't know; lyrebirds are brilliant mimics. But all I heard was their natural voice, a clear, open-throated cry, oddly vulnerable.

The trees on every side were gigantic alpine ash. Shreds of bark dangled in long strips to the ground or festooned the telephone lines that ran along the few yards between the forest edge and the road. The bark strips, as someone had remarked, were just like fuses waiting to carry a flame from a carelessly discarded cigarette up into the canopy of leaves.

Belgrave. It was south of here the fire had started. Suddenly I was on the edge of the disaster. A crew of telephone engineers worked to replace a burned-out pole, blocking half the road to do so. It was the first sign I'd seen of the State Electricity Commission, whose responsibility for Ash Wednesday's fires was to become a subject of national debate.

The road wound out of Belgrave and I was back in middle-class tranquillity: restaurants, craft shops, houses half obscured by thick greenery. This was the environment the people and councils of the Dandenongs had fought to protect. If the fire had burned north rather than south I would have been driving through utter desolation now – the desolation I had seen years before in the Blue Mountains outside Sydney, where it was the mountain ridges that suffered the worst fire damage, flames funnelling up the slope to blast off the high ground everything that would burn.

The houses thinned out as I got higher. There were farms here and a few orchards. Beyond Belgrave and Emerald it became less stockbroker belt, more weekend cottage country. The road got narrower, a road that on Ash Wednesday night had been choked with refugees and rescue vehicles in a hopeless tangle. From this twisting blacktop, hardly wide enough for two cars to pass, returning residents of Cockatoo stopped by police had taken to dirt tracks in a desperate bid to reach their homes.

By mid-morning I topped the hill leading down into Cockatoo.

The road clung to one side of a valley. A few yards down into the dip came the first sign of the fires. On either side of the road trees and grass were burned – probably from a spot fire because it was to my right and below where the valley angled to the south, that the real damage was done.

Down to my right on the valley-floor I could see signs of Ash Wednesday. Heavy wheels had rutted and torn the black soil. Directly opposite where I'd paused the grey-green forest that clothed the hillside was scarred by a long tongue of brown. Every trunk in this zone stood straight and black against pale, bare earth. Each canopy was crisped a uniform brown.

On Ash Wednesday night before the wind change this

valley was empty of fire, leaving the main Melbourne road open. Only a glow to the west in the smoke-filled sky marked where the main fire ate its way south towards the second blaze also burning west of the town.

When the wind change came both fires leaped the hill in a few seconds. Burning branches flying ahead of this one would have lit the canopy at which I was looking now, a path to carry the fire down into Cockatoo, hidden in the fold ahead of me.

I drove down now towards the town, imagining how it must have been that night. How the volunteer brigades roared down in their trucks to try and head off the fire storm rolling over the hill towards their homes, despairing as the flames raced through the treetops, impossible to reach let alone quench.

All along the western side of Cockatoo the same tragedy was being repeated as the fire washed over the hill, flowed down the slope, raced across the town and leaped up the far side in a few moments of terror and fury.

At the bottom of the valley, the road crossed Cockatoo Creek, a tiny stream under a concrete culvert. Just beyond was a road sign – the figure '60' on a disc of metal announcing the speed limit. One half was burned black.

A few yards on another sign, also metal but oblong on two metal uprights, announced I was entering the shire of Pakenham. At first glance this sign seemed less damaged by fire. There was no charring and the wording was pale but readable. I had to get out of the car and examine it to see that the whole painted surface had been thinned by heat, the enamelled message fumed off by the fire storm.

Cockatoo's main street lay to my right. I drove slowly up the slope alone the eastern valley-wall. The town's centre – its shops, the child care centre that had saved so many lives – lay in the hollow to my right.

From the road I could look down the long 'fire corridors' through which the flames had been carried across Cockatoo. The trees in these corridors were cinder-black. Unlike the trees I'd seen coming into town, these had no crisped canopies of leaves. The branches were bare, blasted clean by the superheated air that had swept across the valley that night.

The ground under them was brown-grey ash, the powdered remains of grass, undergrowth, clothing, papers, car tyres, domestic pets; anything combustible. I got out of the car. My feet sank into this earth to the depth of an inch. Between my fingers it felt granular, crumbling – burned out.

Down one such corridor I caught sight of a few green tents, all that remained of the 'city' set up to house and feed the homeless. Somewhere a bulldozer ground, checked and ground again with a roar of diesel. Otherwise there was no sign of life.

At the next corner a huge tree had been uprooted. The earth around it was gouged and torn. Of the house that had stood behind it there was no sign. Not even the foundations remained. Just a pile of ash.

If evidence was needed that this was a fire more terrifying than any other, not a simple bushfire but a fire storm holocaust, it stood on a corner at the heart of the burn zone.

The road sign had originally carried six direction markers on a galvanised steel pole about 2 inches in diameter. The signs were of ⅛-inch metal, bent double. The pole was still erect but the signs themselves drooped now, dangling like dying tulips. They were uncharred and unblackened, just softened, as if thrust for a few moments into an intolerable furnace heat.

Further up the hill I found the bulldozer. It was clearing all that remained of a house on the downside of the road. Above the road, in what must have been the most intense part of the heat zone, a woman was raking the blackened fringe of a garden in front of a two-storey stucco villa apparently untouched by fire.

There were signs everywhere. 'Down But Not Out.' Phone numbers. Addresses where the people who lived in what were once houses could now be found. However attenuated Cockatoo still existed. Disaster revealed what sociologists had always know; a community is less a particular location or locality than a sense of place.

At the top of the hill a young man was piling scrap metal and junk by the road. The heap's foundation was corrugated iron, wrung and twisted, its grey sheen burned to an ashy grey. I recognised the lump he threw on top as the guts of a washing machine, the drum crumpled in on itself.

He caught my eye. A few steps down the hill, by another pile of ash that had been a house, what must have been father and brothers straightened up.

'Go away!' one called to me. 'Yair,' said the boy at the fence. 'Go back home.'

# TWENTY-ONE

## *Blame*

IT'S SAID ironically that there are six stages in any human enterprise: wild enthusiasm, total chaos, utter despair, a search for the guilty, the persecution of the innocent and the rewarding and promotion of the incompetent.

In the aftermath of Ash Wednesday most of these responses would be encountered both amongst those struggling to recover from the tragedy and those who tried to help. It may be that all disasters carry with them a cargo of mismanagement, callousness and thoughtless planning, but the fires of February 1983 had all in quantity.

It was into the category of 'a search for the guilty' that many of the most ironic and astonishing responses fell.

Many people whose property survived Ash Wednesday experienced, paradoxically, feelings not of relief but remorse. An Adelaide housewife said to her neighbour, 'I feel so guilty, looking across to your burned house, knowing mine was saved.' Some kept their blinds drawn rather than contemplate what they felt to be unearned survival. One Cockatoo woman said she was ashamed to put out her washing: many neighbours did not have a washing machine. Some lacked even running water.

The owner of a Macedon house that survived because it had been built on concrete base set into the hillside told a reporter: 'We feel sort of guilty that our house has remained. When we go into the village and welfare workers offer us tins of food we have to refuse.'

The need to conform, to be just like the family next door, even in sharing the effects of a disaster, runs deep in the Australian character. To have survived Ash Wednesday unscathed became for many a source of embarrassment.

Among the first relief workers to move into burned-out areas were psychiatrists and counsellors. The South Australian government retained Professor Beverly Raphael from the University of Newcastle to advise relief teams. 'In these early days,' she told the *Sydney Morning Herald*, 'people are numb. The full horror of it doesn't hit them straight away. The enormity of the loss is really just beginning to strike now.'

Colin and Joy Cook were among the Cockatoo residents forced to live in caravans after the fire destroyed almost all their possessions. Cook, a builder, had salvaged from the ashes of his home only those objects that would not burn : sporting trophies, antique coins and tools. His torment numbed by sedatives, Cook explained to a reporter, 'Before those tablets, I'd just walk up and down the street. In eleven days I got only eighteen hours' sleep. But I haven't cracked yet . . . I came home one day to find Joy in the caravan screaming and ranting. She wouldn't let me near her and couldn't stand me touching her.'

Dr Raphael spoke too of the 'second disaster' – the stress of realisation, often more deeply damaging than initial shock. 'One of the worst things we've found in a variety of disasters,' she went on, 'is that people often feel outsiders are coming in and telling them what to do.'

Environmental experts who moved in to supervise the Aireys Inlet clean-up ran headlong into just such resentment. The animosity against 'out-of-towners' surfaced at a town meeting early in March, when the chairman of the 'Natural Resources Conservation League' and a colleague from Monash University received a rough ride, equivalent to that meted out in the Dandenongs, where the environmental prejudice against tree-cutting was blamed by some for the fires' swift spread.

A week after the fires the local nurse at Aireys Inlet told Richard Guilliatt:

It's now and for the next few weeks that we are going to have problems. Red Cross has moved out, the Community Welfare Services people are coming and going, and support has really been reduced dramatically.

People are reverting to day-to-day living in completely altered circumstances. People are starting to feel they're being left on their own. The next two weeks I reckon will be the real crisis. We're seeing signs now. People are getting edgy; there's a lot of anger here now.'

This belligerence was far from unique to Australia. There's a universal human need after disaster to find, as John Steinbeck wrote in *The Grapes of Wrath*, 'someone who knows what a shotgun's for'. There was no shortage of scapegoats, imagined or real. In the days after Ash Wednesday most of them were relentlessly taken to task by victims driven beyond their capacity to absorb more psychic punishment.

Adelaide CFS Fire Chief, Professor Peter Schwerdtfeger, stood in the ruins of his century-old Adelaide Hills home and preached a final solution to the problems of the hills, whose destruction he blamed on the new residents.

Pay these people for their losses and tell them to get the hell out of the Hills [he told the *Sunday Mail*]. These hills have to be returned to national parkland. It's the only way we'll prevent future fires of such magnitude.

People and machines bring fire. You can't say the thunder of heaven opened up and started all this. It was people, and they're criminals.

The greatest rage, however, and a large share of the blame for at least the fires that destroyed Macedon and

145

Mount Macedon, was directed at the State Electricity Commission.

Very few of Australia's electrical cables are underground and the outmoded system of insulated cables and wooden poles invited disaster in a country where the climate swings between baking heat, desiccating wind and soaking downpour. Plastic insulation on the cables had quickly cracked and peeled off or been rubbed away where, as was often the case in country areas, power poles ran along tree-lined roadsides. The poles themselves, some thirty years old, rotted in the ground and often splintered and fell without warning. SEC teams travelled the state, propping up rickety poles with braces and ropes and marking those in need of replacement with the red cross of condemnation. Such replacement was notoriously long in coming.

The commission had no illusion about the risk to the environment of faulty electrical equipment. Television journalists after the Ash Wednesday fires unearthed a documentary film called *Cutting the Fire Risk* produced by the SEC itself in which technicians demonstrated the cutting ability of a naked power line. It ate through a 6-inch branch like a saw.

· Country residents had worse stories. The wind slapping a cable against a tree could send sparks cascading into dry grass beneath. A toppling power pole would do the same. Frequently poles ran through rows of trees positively inviting a fire.

Extensive reparations had already been paid by the commission to farmers burned out by fires which it had been accused of starting. Between 1969 and 1983 it settled claims for a staggering $30 million. $20 million of that was paid out in 1977 for fires near Streatham and Cudgee, areas ravaged on Ash Wednesday as well. The embers of Ash Wednesday were hardly quenched when Victorians and South Australians in the fire-devastated areas blamed faulty electrical wiring for many of the worst fires.

People attempting to drive through the Macedon area

the day after the fire found the roads blocked, sometimes by police but more often by citizens' groups. There were accusations that the SEC, 'with the knowledge and consent of the police', had attempted to remove evidence of faulty wiring in the area around Macedon: within weeks of Ash Wednesday local residents would institute a $200 million compensation suit against the SEC.

David Munro, appearing before the Victorian Supreme Court on behalf of a group of Macedon and Mount Macedon residents, aired some alarming allegations. As the *Australian* reported the hearing:

Mr Munro told the court a fire had started at the same place a year ago, but all the SEC had done was to install one spreader to hold parallel lines. He said the spreader had been wrongly placed.

He tendered photographs which he said showed old rub marks and a burn mark on a tree from a power line three days after the latest fire.

Mr Munro said a farmer had seen SEC employees taking pieces of bark from the tree. Another witness inspected the tree and later it was obvious that the burn marks had been cut out.

Mr Munro said he had evidence that a voice over the SEC's radio at the time of the fire said, "We've done it again. Get out of here in a hurry. It's our wires".

An order restraining the SEC from interfering with potential evidence was handed down, varied later when the judge acknowledged the commission had the right and duty to safeguard both the electricity supply and any evidence of possible negligence. He declared accusations of negligence and interference against the SEC irrelevant. But these charges were given wide publicity both on radio and television and in print.

The Australian edition of the investigative TV programme *60 Minutes* dug into the claims in a segment called

'When the Sparks Fly', and came up with some extra-ordinary and damaging material. 'On bad fire days,' said one man interviewed, 'the SEC is the biggest lighter of fires in the state of Victoria.' Local firefighters toured fire-devastated areas with the cameras, showing the crooked, condemned power poles, the points where cables have rubbed raw against untrimmed trees.

*60 Minutes* showed that the SEC might have played a major role in the Ash Wednesday disaster. There was footage of the field and the tree where the Macedon fires had started. It was, claimed those interviewed, a notorious danger spot which had burned before when cables slapped against a tree or clashed together, sending sparks showering into a paddock.

The film showed power lines running through groves of cypress trees where the wind had worn away most of the cables' insulation. Immediately after Ash Wednesday, it was claimed, the SEC bulldozed one row of trees and cut a clearway through the other for its cables. But by then the damage had been done.

In South Australia the same accusations were levelled at power cables. In May a coroner's inquest would be convened by a broken power line near Naraweena in the south-eastern section of the state where the worst fire in that part of South Australia is believed to have begun. The coroner examined the rotted truck and limbs of a swamp gum that were marked with 'charring and abrasions' and the fallen power pole nearby. A QC retained by local residents asked the coroner to note that the gum was the only tree close to power lines in that paddock.

It is beyond refutation that Australia's mis-handling of electricity contributed immeasurably to the Ash Wednesday disaster. Elementary precautions against fires, including the imposition of a statutory cleared zone along road verges near power lines, have been consistently ignored over the years, even when other countries like New Zealand and the United States demonstrated their efficacy.

If a carelessly maintained wire or rotted pole started a bushfire, it merely completed a process already begun decades before. There was the weariness of ritual about the claims made against the SEC in Victoria and its response to them. Far more shocking and more unexpected among the charges of culpability laid in the days following Ash Wednesday were those of arson. Within days it became clear that many of the fires, far from being accidental, had been deliberately and maliciously lit.

# TWENTY-TWO

## *Give till it hurts*

AFTER THE disastrous fires of February 1851 the citizens of a shocked Melbourne subscribed £3171 to a fund for the homeless. A committee of prominent philanthropists and churchmen was set up to administer what was, for the time, a large sum. The disbursement of the money, however, led to one of the worst scandals in the not-unsullied history of Australian charity.

Ten people died in the fires. Thousands lost their homes, many more their stock and crops. For weeks after the fires ruined farmers walked into Melbourne from hundreds of miles away to collect a new pair of trousers, boots and clothing for their families. Then they turned and began the long, weary trudge back to the charred paddocks of their smallholdings.

It was obvious to everyone in the city that these fires had been the worst in the nation's history. Yet the charity committee set up to distribute the publicly subscribed funds announced that 'because the damage supposedly suffered fell considerably short of original estimates', only half the money would go to victims. The remainder was split up among the charities closest to the hearts of those on the panel.

Furious subscribers convened a protest meeting. According to one report, 'Thousands listened in disgust as clergymen of rival Church of England and Presbyterian denominations accused one another of misappropriating public funds.'

Spokesmen for the subscribers described the plight of a farmer named McLelland from Plenty River, 35 miles north of Melbourne. McLelland lost six members of his family in the fires, as well as his home and 1100 sheep. In an attempt to rescue his children he suffered severe burns. He was discovered, his drover said, 'writhing and groaning in the creek with one of his arms no more than a burned stump'. For his agony and loss the fund offered McLelland £40.

The Ash Wednesday fires of 1983 produced an wellspring of public generosity no less unstinting than that of 1851. It was matched by the federal government, which on February 18th announced an immediate grant of $10 million to Victoria and $6 million to South Australia. Private donations to the victims, ranging from gifts of food, clothing and blankets to cash, collected in hundreds of informal appeals, as well as those on national radio and TV was estimated at a further $5 million.

The artistic community rallied with characteristic generosity. By February 27th an afternoon benefit concert had been organised at the Sydney Opera House. Supporters paid up to $20 a seat to hear some distinguished performers, including Dame Joan Sutherland, who delayed a scheduled return to Switzerland to perform with stars of the Australian Opera. The sold-out concert raised $37,000 for fire relief.

Across town, radio station 2SM held a 'Rock for a Reason' concert for 5000, raising $25,000. Theatres held benefit performances of their plays. Sculptor Peter Latona, showing a series of bronzes at the Opera House Exhibition Hall, including a bust of the tragically dispossessed Dame Joan Hammond, volunteered to donate any proceeds from its sale to the fire victims.

Some schemes offered less tangible but still valuable assistance. A plan was announced to treat the families of those dead in the fires to holidays in the southern New South Wales resort of Merimbula. And the owners of a

12-metre yacht challenging for the America's Cup took the children of incinerated Cockatoo on a well-publicised trip around Port Philip Bay.

Public response to appeals for help, either in money or kind, was almost embarrassing in its prodigality. A letter of March 7th from Dame Phyllis Frost, chairperson of the Victorian State Relief Committee, spelled out its dimensions. After acknowledging that, 'the hours of work put in by thousands of volunteers over the past ten days would be hard to estimate', she noted:

All food, clothing, furniture and other goods sent to the committee have been sorted, packed, labelled and many already distributed.

Any dirty or torn clothing has been baled and will be sold by weight as rags. Metal articles such as pots and pans with holes will be sold as scrap metal . . . We are now holding over 100,000 cubic feet of good useable clothing as well as other materials. There are approximately 7000 to 8000 people affected by the fires so this would mean that, because of the generosity of people all over Australia and overseas, we could supply 20 cubic feet of clothing to each person, which is more than the average person would use in 10 lifetimes.

In a final plea Dame Phyllis announced: 'We have not wasted any articles but have stored them safely, moth and rodent-free. If there should be any major disasters in Australia (God forbid) or in any other parts of the world, we will be happy to immediately forward those splendid articles of clothing, sorted and ready for use.'

As in previous disasters, the insurance companies showed less enthusiasm for paying claims than they had for collecting premiums. The homeless were warned to photograph what remained of their houses; they would need such evidence to convince insurance assessors of replacement value.

Industry spokesmen hinted at the possibility that the insurance industry would need massive and worldwide re-organisation to cope with the expected avalanche of claims. All major Australian insurance companies had taken the precaution of re-insuring themselves against massive losses like those of Ash Wednesday. As the first claims began to reach the speedily created Insurance Emergency Service, the companies sent out demands against their own catastrophe insurance placed around the world.

One large international insurer, American International Underwriters, hinted darkly to the *Australian Financial Review* that these claims could precipitate a 'major shake-out in world insurance. Heavy competition in the insurance business had led to premium-cutting; losses anticipated from the depredations of El Nino – including the Ash Wednesday fires and damage caused by rain storms and flooding on the American west coast – might, hinted a senior vice-president, 'shake out a few of the irresponsible underwriters'.

Inside the industry a few people speculated on what might happen if the fires were followed by some further catastrophe that placed an additional strain on the be-leaguered insurance industry. There was an ominous tendency for disasters to come in pairs or even threes. Since 1971 insurers and the federal government had been dickering over a national emergency insurance scheme supported by the federal government and the insurance industry. Like all disaster measures, however, it seemed irrelevant during times of peace. In 1979 discussions were suspended.

In February 1983, unaware that the next natural disaster was already waiting in the wings, the insurance companies swung into their pre-arranged programme for emergency claims' processing. Over the next fifteen days, Melbourne's IES office processed 4000 claims to a total of $138 million. Claims against the Adelaide office were estimated at 1000, worth $48 million.

The Victorian government's decision to issue an immediate $1000 to every homeless family appeared a prompt and intelligent response to the disaster. But it was not long before protests from the dispossessed were heard all over Australia.

Some complaints hinged on the suggestion that those who received emergency aid might be required to pay it back from insurance claims or cash received from other relief agencies. By late March the Victorian government would confer trustee status on the members of the main bushfire committee. 'This will formally allow the committee,' said one report, 'to make arrangements with fire victims to recover money if they eventually receive compensation through insurance or other legal action.' To many it seemed a harsh and unfeeling ultimatum delivered to people who, living in tents or squeezed into the homes of friends, were only then beginning to come to terms with their loss.

The inevitable false claims against bushfire relief, swiftly dealt with by the courts, recoiled on the government. Instead of demonstrating that relief funds were effectively controlled, the widely publicised convictions soured public generosity. Such reports must have contributed to the failure of many individuals to honour their initial pledges of help. There was a growing sense that much of the money would be, if not misappropriated, then used to benefit the wrong people.

From a letter published by the *Sydney Morning Herald* on February 22nd it is possible to gauge a growing public cynicism at the disbursement of relief. The writer had lived through the December 1968 fires in the Blue Mountains, and made some charges which, if true, appeared to offer a lesson in 1983.

I would like to point out [the letter began] a few vital points to anyone who is thinking of sending donations . . . A panel is usually set up to collect donations. After a long time, in which arguments break out among the so-called panel of advisers and helpers themselves, the moneys are distributed in a very discriminative way . . .

If you are not insured fully, you will have to hound the distributors of the moneys so freely and lovingly given. The Leura and Blue Mountains fires were one of the worst disasters known . . . Yet the money lay there, a great deal of it never distributed.

By March 1983 a scandal was brewing – one of the many that would follow the fires.

The Melbourne radio station 3AW withheld $1.75 million contributed by its listeners when it learned that the government had handed out only $2 million of the $9.8 million in its relief fund. According to the station's general manager, quoted by the *Australian*, 'Many people who gave to the station's appeal had expressed concern after learning the money would be placed in the Victorian Treasury Department Appeal Trust.' The report went on to announce that 'The 3AW allocation of funds would now be determined by the Salvation Army, the Country Fire Authority and now the Red Cross in conjunction with the radio management.'

By mid-March complaints about the distribution of relief reached Victoria's state parliament. Opposition leader Jeff Kennett was quick to echo this popular dissatisfaction. 'To date we don't know where the money is, whether it's been invested independent of government funds, what the guidelines are for distribution.' He accused the government of using the money as a 'slush fund'. 'How long,' he asked, 'is this money, or part of it, to remain in the cash management account, propping up the deficit, before being given to the fire victims?' He described the entire bushfire relief programme as 'one huge bureaucratic bungle'.

Ash Wednesday had more widespread political repercussions than those in Victoria. The campaign of incumbent Malcolm Fraser to defend his administration against a strong left-wing assault from the Labour Party suffered immeasurably from the loss of news coverage.

Fraser had many friends in the bushfires areas, both around Gisborne and in his electorate of Wannon in western Victoria. He spent some time trying to establish whether they had survived the fires, then cancelled two days of electioneering appearances to visit the fire areas. When a reporter pressed him to comment about the dispute over using Hercules MAFS planes and who should foot the bill, Fraser snapped, 'This is no time to quibble about money – people are dying.'

After the election Liberal Party president Tony Eggleton was to acknowledge that the fires fatally interrupted public interest in the election and its issues. By the time voters began once more to take notice of politics, Fraser's campaign had run out of steam while that of Hawke was at its peak. Labour swept in with a convincing majority a month later.

Victorian Premier Cain did his best to defuse the problem on March 17th by detailing the distribution of funds to that date. But he was forced to acknowledge that only $2 million had been disbursed out of the available $9.8 million – itself only half of $18 million in public pledges. For the first time he quantified the loss from Ash Wednesday's fires – a staggering $190 million, made up of $164 million in private property, $16 million in state property and $10 million belonging to local and public-sector assets.

Political writer Paul Lynch described the struggle between Cain and Kennett over Ash Wednesday as, 'a shameful display of political bickering, point-scoring and slurs'.

There is obviously a tradition in Australia, however, for national disaster to be exploited more for the benefit of those untouched by misfortune than for the victims. Cain was forced to agree publicly that there had been 'mistakes which must be freely acknowledged'. 'There is no doubt, too,' he said, 'that further lessons will be learnt from the experience of this fire season and of Ash Wednesday in particular.' It was the first major statement in what was to become the next stage in the Ash Wednesday disaster. Who had burned Australia? Who was to take the blame?

# TWENTY-THREE

## *The end of Ash Wednesday*

ALMOST TWO months after Ash Wednesday an enquirer wrote for guidance to Dr Lyn Barrow, the psychologist whose column of advice appears in the Sydney *Sun Herald*.

What sort of person would deliberately set fires in the bush knowing, as they must, the horrible destruction that this will cause? I find it unbelievable that anyone could do this. Yet incredible though it seems, both the police and the fire authorities have stated that some of our recent and most catastrophic fires were deliberately lit.

How could any normal human be so completely unmindful of the destruction and suffering that their actions cause?

Dr Barrow couldn't offer much help. He did point out that the word 'arson', then being thrown around a great deal in the press, was inaccurate; arson is setting fires for profit, not out of malice or for some obscure private satisfaction.

He used the accurate term for the fire-starters – 'pyromaniacs' – and noted that:

They are not insane in the legal sense and may be quite normal in other aspects of their lives except for this disorder – the compulsive urge to set fires and watch them burn.

Some say the pyromaniac is a sadist, a destructive and

unconscionable person with fire fantasies. Psycho-analysts state that the fire-bug experiences marked sexual excitement while watching the flames.

But, asked some people, isn't setting fires simply a natural response to the volatile Australian environment? The Australian farmer and householder burns his property as a yearly ritual of husbandry; the barbecue has become Australia's most popular social event and a lively source of accidental bushfires; and, particularly in the Sydney area, the malicious and often calculated setting of fires has reached the proportions of a crime wave.

Throughout the 1980s fires at schools became a regular feature of the weekend's crime statistics. In the city's Western Suburbs, the flat dormitory area that stretches beyond the luxurious harbourside areas almost to the Blue Mountains, disaffected and bored juveniles routinely took out their resentment on school buildings, breaking in first to vandalise then, as they left, setting fire to the classrooms both to cover their tracks and commit a final act of defiance.

During 1983 this pattern of destruction entered a new and, at least for the city's affluent middle class, more alarming phase when a wave of car burnings spread across the city from the Western Suburbs to the high-rent seaside districts. In the first six months of the year sixty-five cars, six trucks, a motor-cycle and a boat were burned. Damage was estimated at $2 million.

The police called them 'copycat crimes'. A spokesman commented, 'In nearly all cases, nothing has been taken from the cars before they are burned out. They just like to see their work on television.'

The fire-raisers seldom robbed the cars. They merely chose a likely vehicle standing before some high-rise block, doused it with petrol and lit a match.

There was an insouciant and egalitarian style to the act that the witless found attractive. The day after Sydney papers carried news of the phenomenon, a Brisbane paper picked up the news. That night, three cars burned in Queensland.

Setting fires in Australia has none of the majesty of Europe, where a field in flames marks it presence for miles with a plume of smoke, and farm-workers maintain a healthy respect for the possible dangers of a runaway blaze. The Australian environment is 'out to get you' no matter what you do. Fire is another random and destructive element, like the voracious kangaroo and rabbit, the floods and droughts and the insect pests. You taunt it, daring it to destroy you.

Reports on how fires start often have an air of resigned hilarity – like this one about a fire in the great dry of February 1975 (another El Nino year):

> Rabbits have been a curse for Australian farmers ever since some silly man introduced them to the country last century as pets, but starting bush fires seems to be a new rabbit hazard.
>
> In February farmers were doing some burning off on the property of Mr J. S. McKenzie of Port Germein on the Spencer Gulf in South Australia. A rabbit caught fire, dashed across the fire break and did an incendiary 'streak' through an adjoining paddock.
>
> The fire raced quickly up the side of Flinders Range and burnt out thousands of hectares of country. One hundred firefighters turned out, but the fire got right out of control and at one stage was burning on a 16 kilometre front with flames up to 7 metres high. The blaze was clearly visible from Whyalla nearly 32 kilometres across the bay.

It seems extraordinary that anyone would light an open fire in February in Australia. Yet farmers and picnickers routinely do so. As do those people who achieve some crazy thrill from watching the ulcer of flame eat its way outwards through the dry grass.

161

After Ash Wednesday, Keith Johnson, the Victorian CFS's Deputy Chief Officer (Operations) told the *Bulletin* that there had been outbreaks of bushfires in Victoria since November. 'There's no doubt in my mind that the majority of these fires have been deliberately lit,' he said. 'We have put a lot of time into reducing the causes of fire but we can't reduce the number of deliberately lit fires.'

The first charges of fire-setting on Ash Wednesday came from Adelaide. On February 18th Andrew Mervyn Davey was charged with lighting a fire in scrub at Kersbrook and remanded in custody 'for his safety' at Para Hills Local Court. Magistrate Mr A. Moss, SM, said: 'I must look at the safety of the public, and in this case, the safety of the defendant. He will not be the most popular person in Adelaide to the public.' Seven policemen guarded Davey as he left the court.

The *Adelaide Advertiser* report noted that:

Davey, 19, unemployed sheet metal worker of Adelaide Rd., Gawler, was charged with unlawfully and maliciously setting fire to scrub at Kersbrook on Wednesday.

He entered no plea and was remanded in custody to appear in Para Hills Local Court on March 4.

Mr R. W. Winter, Duty Solicitor for Davey, said he had advised Davey to apply for a suppression [of his name] because of the general atmosphere of the state, but Davey had instructed him he had nothing to hide and felt he would be found not guilty.

The week after Ash Wednesday, as fires again erupted through Victoria, police sent out aircraft and helicopters in a search-and-deter mission against fire-starters. The head of police operations had warned routine patrols to be on the look-out for anything suspicious. 'The first whisper of smoke and my people will know and contact the suitable authorities,' he said.

On March 4th the CFS reported that a fire raging in the

Gippsland area of Victoria had probably been deliberately lit. There was, said police, 'a lot of evidence' to suggest it had been set by fire-starters.

Then on March 22nd in Adelaide, the most serious charges yet were laid against an alleged 'arsonist'. Darren Mark Bing, 20 years old and unemployed – almost a pattern amongst Sydney's car-and school-burners – faced charges that he had unlawfully and maliciously set fire to grass at Tea Tree Gully on February 16th. To this was added a charge of having murdered Betty Jean Coventry at Lenswood and unlawfully and maliciously inflicting grievous bodily harm on Phillip Geoffrey Williams at Houghton, both people being victims of Ash Wednesday's fires.

At an earlier hearing Bing was also charged with setting fire to the same area on November 8th of the previous year and, the day after Ash Wednesday, unlawfully burning grass and attempting to set fire to the First Hope Valley Scout Troop building.

The most extraordinary revelation came on March 25th in Melbourne.

After being questioned for most of the day a member of the Melbourne Metropolitan Fire Brigade, formerly a resident of Cockatoo, was charged with lighting the fires that destroyed that town and with setting a similar fire exactly a year before. He also faced two further charges of lighting fires on a day of total fire ban.

After the horrors of February it was easy to assume that the fire risk was past. Brigades battling the Gippsland blaze were, however, as helpless in stemming this fire as they had been a month before. 'We're facing the same problems we had with the other major fire here,' said a spokesman for the commission. 'It will burn right through the night without any problem because the area is so dry and there's not enough water in the creeks to stop it.'

Fires continued to flare all over Victoria. Near Mallacoota, close to the New South Wales border and the scene of another fire in January 1983, a major blaze broke out near the site of the first. Forests Commission firefighters accused 'arsonists' of setting it.

March 9th 1983 was Sydney's hottest March day on record; the heat rose to a choking 39.8°C and the evening Southerly Buster reduced it by only 13°. Around 9 p.m. Sydneysiders experienced a terrifying reminder of the hours before the Dandenongs burned.

Through the windows, opened to let in what remained of the cool change, came the smell of burning eucalyptus. A dusky red shroud settled out of the dark over Sydney. A 'crawl' on TV sets warned residents not to make further phone calls to the emergency services; there was no risk of Sydney burning. 'It's driving us crazy, said a spokesman for the Sydney Fire Brigade. 'Smoke from various bushfires has caused a rash amount of calls.'

Some of the smoke came from a fire burning in the Lane Cove National Park, north of Sydney but most was carried north from fires on the Victorian border. It was as if the fires, just before their rule expired, were reaching out with a last warning of their power.

This was the high point of the Ash Wednesday holocaust. In Victoria and South Australia the fires of the month before remained quenched and though new blazes did spring up no disastrous wind change intervened to turn them into fire storms. Though man feared it the Blue Mountains did not burn. Fires around Sydney were confined to minor blazes with no loss of life. But the real disaster was not over.

Within a few days of the smoke cloud over Sydney, their domination of the Australian continent would be supplanted by another old adversary of man's – utterly different from the fires but no less destructive. It was time for El Nino's last ironic intervention.

## TWENTY-FOUR

### *The end of El Nino*

To say that all of Australia prayed for rain in March 1983 is vastly to minimise what had become, after the Ash Wednesday fires, a psychic need. Australia seemed to be desiccating, crumbling, breaking up before the eyes of its horrified inhabitants, who could do nothing but watch helplessly, like mourners at the death-bed of a relative terminally ill.

By now the part played by El Nino in February's ecological disaster was widely accepted. Australia had belatedly become aware of just how intimately its way of life was involved with that of the Pacific ecosystem. Details of El Nino's ravages on the far side of the world were filtering through. Ecuador had 108 millimetres of rain in November 1982 as against the more common 8 millimetres! and in January an incredible 614 millimetres. London's *New Scientist* recorded 'widespread flooding, loss of life, property damage and economic disruption across one-third of Ecuador'. Not only did El Nino show no sign of abating in the early months of 1983, its effects were also seen to be spreading to the tropical Atlantic, whose waters also started to warm up.

The hurricanes that struck the Caribbean and western United States in the last days of February drove the point home. The Californian storms killed thirteen, injured fifty and left 6500 homeless. A Cuban hurricane inflicted damage estimated at $650 million and, closer to home, Fiji lost

165

seven lives to Cyclone Oscar in the same week. Hawaii's Kilauea Volcano erupted – one more note in a symphony of climatic discord.

Australia at last started to take notice and to view the drought and Ash Wednesday fires not as narrow local concerns but as aspects of a global catastrophe. There was a notable increase in the reportage of climatological anomalies. Feature articles covered every aspect of the crisis from the long-term effects of the slash and burn farming to the threat to Australia's wine industry posed by the droughts and fires. Regional papers that had ignored the syndicated reports from the *New York Times* and the *New Scientist* three months before belatedly featured news of El Nino and its influence. Climatologists and meteorologists, traditionally among the forgotten men of science in Australia, found their opinons sought and valued.

One of these was Barrie Hunt of the Australian Numerical Meteorology Research Centre. In March 1983 his eight-person research group was four years into a ten-year computer simulation of the Australian atmosphere, part of a project which aimed, as he told the *Australian*, to 'look at the whole complex pattern and the interrelationships between ocean, atmosphere and the winds'.

Like David Packham at the Chisholm Institute, whose warnings about bushfire risk had proved so dismally prophetic, Hunt's work was undervalued by the scientific establishment and risked never reaching its conclusion.

According to *Australian* Science Correspondent, Jane Ford, the group faced amalgamation with a new CSIRO atmospheric unit; one member had already left because of the work's uncertain future. Hunt hoped to find Reserve Bank funding to keep the unit together. 'I think the work will survive,' he said optimistically. 'It is well recognised, but we are going through a hiatus at present and the project is gradually running down.'

On March 4 the *Australian*, under the headline 'The Weather Watchers Wait With Uncertainty', published a sober report on Hunt's work. For the first time the Australian media treated El Nino's influence as an accepted fact rather than fantastic speculation.

The mass of warm water ruling the Pacific's weather was now, Hunt said, 'king size', a giant pool many metres deep stretching from just off the coast of Ecuador to the tropical east coast of Australia. As long as it remained in place, Australia's weather would continue to be ruled by the high-pressure zone that diverted water-bearing winds. But Hunt was hopeful. Measurements of the water temperature in December showed it to be 5°C above normal but by February it had cooled two points. 'As long as nothing happens to disturb the reversal,' he said, 'we have quite an encouraging drought-breaking pattern. But anything might happen and the present cooling down of the ocean mass could stop. Without this cooling the drought won't break.'

Jane Ford spelled out succinctly what Australia could hope for when the trade winds began blowing once more. 'The warm tropical ocean pool should cool and the circulation be reversed. Warmer water could then flow down the east coast of Australia, creating rain producing conditions and the normal rainfall patterns would return.'

The article was remarkable for its timeliness and marred only by an understandable caution about the effects of El Nino's reversal. Even as the paper went to press, 'rain producing conditions' had appeared as the temperature of the Pacific surface water dropped to a level where the trade winds could once again move moisture towards the parched Australian continent.

The cautious hoped for rain. What they got was Cyclone Elinor and Cyclone Ken.

On March 2nd and 3rd, 1983, the high-pressure system that had rested like a shield over Australia for more than a year slipped away to the north. Into the vacuum surged two vast masses of hot, wet air from the north-east and north-west, one sliding in over the arid and lightly populated north-west

167

coast of Western Australia, the other driving straight at Queensland's highly developed Sunshine Coast.

As they moved in both masses interacted with cells of low pressure to create swirling storm systems. Ken turned into a rain depression and poured water on to the parched deserts of Western Australia. Elinor spun off a litter of thunderstorms that drove south across Queensland, New South Wales, even into Victoria. Rain fell in areas of northwestern New South Wales officially designated as droughtafflicted for more than three years. In some areas the first few millimetres simply sank into the parched earth and graziers saw the change as just one more taunt thrown at them by nature. As one commented. 'Our tanks are still dry. Some of the creeks are running for the first time in a year, but they'll dry up in a few days and we'll be as desperate as ever for water.'

He was being excessively pessimistic. The drought was thoroughly broken. But, as Australians had learned over centuries to expect, it had done so with a vengeance. Australia was about to experience the conditions endured over the last year by Ecuador and California at the other end of the El Nino cycle.

Where the two systems collided over the Northern Territory and South Australia there was freak weather: there were hailstones the size of golfballs at Truro, elsewhere simply a deluge, a seemingly endless torrent of warm rain.

'The state's gone crazy,' said an astonished South Australian CFA man called out at one-thirty in the morning with 200 other firefighters, emergency service personnel and volunteers to rescue stranded residents from the Barossa Valley. 'Most of our guys have hardly recovered from the fires, two weeks ago, and now we are almost up to our necks in mud.'

It was mud not water that inundated the vineyards and villages of South Australia. Denuded of grass by drought and fire, shaken loose by two years of winds, the topsoil mixed with water as readily as if it had been salt. 'It was like a sea of mud surging through the house': the description recurred throughout reports of South Australia's floods. It would be repeated later in other states. Without vegetation to help it soak away the water gushed into rivers whose beds were baked to the texture of brick. In some places water washed over bridges 45 feet above the river-bed, leaving the concrete plastered with brown mud.

Flash floods tore through towns whose residents had tried to escape from the heat and humidity by sleeping out of doors or in tents and caravans. A 15-foot wall of water burst into the town of Gawler, 30 miles north of Adelaide, sweeping up the caravans from its riverside parks and whirling them away. More than a thousand people had to be evacuated. The *Sydney Morning Herald* recounted how nursing sister Marie Gill and a friend were swept into Tanunda Creek by the flood. The water washed Marie helplessly downstream and slammed her into a tree. She was found dazed and semi-naked miles away.

In Western Australia and Queensland, unscathed by the fires, the cyclones did less damage, though small boats off the Queensland coast radioed for help and a hospital on the coast near Yeppoon was evacuated as waves from Elinor battered the coast. In Western Australia aboriginals living on outlying settlements were threatened by the heaviest rainfall in thirty years. Four boats filled with volunteers fought their way to a stranded group through flood waters swirling with shattered tree limbs and the corpses of drowned cattle. As these carcases threatened to contaminate the water supply (as backed-up sewers had in South Australia) the emergency services flew in supplies of water-purifying tablets to Fitzroy Crossing, worst-hit of the outer settlements.

Even the Northern Territory, the most arid of Australia's states, experienced the devastation of flood. It is significant that the first fatality of its flood was a man who had camped in the usually dry bed of the Todd River near Alice Springs. His body was washed 3 miles downstream by the same flash flood that inundated large areas of the unsuspecting town.

Over the middle weeks of the month enough rain fell on Australia to break the drought of five years. From March 18th until the 21st, following in the wake of Cyclone Elinor, a rain-filled depression moved at a ponderous 6 miles an hour south-west across most of Australia's eastern states, drenching the parched ground.

'We've got kids here who have never seen this much rain,' reported one delighted resident of Cunnamulla near the Queensland border. 'Some people have been driving around in it all day. It's beautiful, it's unreal. If it keeps up, it will turn the whole season around.'

It was nice to have something to celebrate after years of hardship. The children playing in the puddles and the elated farmers driving deliriously through the downpour had no thought for the consequence of this reversal, more disastrous in its way than the fires of February.

## TWENTY-FIVE

### *A couple from another world*

THE MOST widely publicised result of the March floods was, ironically, the inconvenience it caused to Prince Charles, Princess Diana and their baby, Prince William, then about to embark on an Australian tour that included the Northern Territory.

Alice Spring's Casino Hotel, extensively promoted as the royal family's home during their outback visit, was up to its first floor in flood waters as soon as the Todd overflowed – waters which also destroyed the road linking the hotel to town. A total of 180 people sheltering there had to be airlifted out by helicopter and the regal home-from-home was relocated at the inelegantly named (but dry) Gap Motel. A newspaper cartoon showed the royal couple perched on a limb a few inches above the floods as Prince Charles observed that he now knew the meaning of the Australian phrase 'to be up a gum tree'.

Charles and Diana made a well-publicised visit to burned-out Cockatoo on a cold Friday at the end of March. Damp, cool winter weather now ruled Victoria – the climate most people expect in the Dandenongs in March. If you forgot you were in Australia, the dark trunks and brown foliage of the seared hillsides might remind you of autumn in the eastern United States. Children wore parkas and scarves as they waited among the 4000 watchers for the royal motorcade to wind its way up from Melbourne.

Bereaved relatives of the Panton Hill firefighters stood in

a nervous, isolated group. surrounded by CFA volunteers in uniform. The families of the Narre Warren brigade who died on the same Beaconsfield road refused to attend. 'I think they felt they had enough reminders,' a CFA spokesman told the *Age*.

Nobody really imagined that what had happened in Cockatoo could be made explicable to any foreign dignitaries, no matter how sympathetic. None the less politicians brought out the maps to show where fire storms had raged, where rolling blue globes of gas had cannoned into houses and set them afire in a fraction of a second.

The impulse behind the visit was generous, the effect almost grotesquely alienating. The CFA presented Charles with a firefighter's yellow hard hat, soon filled with gifts from the children who trailed Diana as if she was a movie star. Posies, an Easter egg, a cellophane-wrapped teddy bear all went into the helmet and police ferried bouquets by the armful back to the official cars.

The pre-school centre, where hundreds had survived the fire was inspected. It had been furnished for the occasion with victims of Ash Wednesday, who made themselves dutifully available for questions. The *Australian* recorded this exchange between Charles, Diana and Sheila Griffith, a bandaged veteran of Ash Wednesday with burns on 40 per cent of her body:

CHARLES: Are your burns painful?

MRS GRIFFITH: Not now.

DIANA: It's wonderful what they can do today with surgery. I've had first-hand experience of this myself.

MRS GRIFFITH: We lost our son.

CHARLES: We're very, very sorry.

DIANA: How wonderful the people of Australia are. It's marvellous how they rally around an emergency. (Pause.) I think a lot of people will want to leave the area now.

MRS GRIFFITH: No way. We're not going to leave.

Later Mrs Griffith told the *Australian*: 'The visit gave us all a great lift in the big job of rebuilding our homes and lives. Diana has brought a lot of sunshine into our unhappiness.'

But the *Age* caught the exact tone of baffled concern when it headlined its report: 'A Couple from Another World Try to Understand the Hell of Cockatoo'. Since no other nation would allow such disasters to take place decade after decade, how could anyone from another country hope to understand?

A researcher looking for evidence that nature in Australia existed only to play a taunting cat and mouse game with man – that it would always, as the farmers feared, 'come back and get you' – will find rich material in the events of March 1983. It was almost as if some of the assumptions embodied in the mythology and religion of the Australian aboriginal were being offered concrete environmental support.

For decades aboriginals had been fighting to retrieve their rights to the land they saw as having been usurped by whites. They argued persuasively that a continent so peopled with ancient ghosts and elemental spirits could not be 'owned' in the normal sense – that land was indivisible, possessed of a mystical life which defied conventional conceptions of exploitation. In 1982 a company mining semi-precious stones on what local aboriginals regarded as sacred land offered to negotiate a deal. It was astonished when the local people, instead of accepting a cash settlement they desperately needed, voted that the company simply return the stones to the hallowed earth.

It is likely that these original Australians in forming this view of the continent merely acknowledged that its complex ecology was too sensitively balanced to be disturbed without dire results. Each attempt by white society and industry to form or shape the land for their own profit has met with disaster as the ecological pendulum swings inexorably between opposites. The next swing began while people reeling from the effects of fire struggled to deal with those of flash floods.

Fences burned out on Ash Wednesday and restrung only days before were torn out by the waters and festooned over the few trees that remained. In the Dandenongs around Clare and in parts of the Barossa Valley, where people were trying to rebuild houses unroofed or partly burned by bushfires, cyclonic winds tore away the flimsy polythene sheets used for interim protection, admitting rain to ruin new carpets, plaster and paintwork. Since many insurance policies specifically excluded flood damage as an 'Act of God', some home owners found to their horror they had survived one assault with resources intact only to be levelled by another.

In southern South Australia loggers trying desperately to fell stands of soft-wood not irretrievably damaged by fire were forced to bulldoze the trimmed logs into lakes to prevent the formation of stains within the wood that would render it unsaleable.

There were plenty of suggestions that a vengeful deity had singled out South Australia in particular for divine retribution. An *Adelaide News* columnist remarked: 'This place is getting like Pharaoh's Egypt. Any day now I expect to hear that we have been smitten by a plague of locusts.' A winemaker in the ruined Barossa Valley north of Adelaide agreed: 'I don't know what the Barossa has been doing to deserve this. All we need for the complete set is the pestilence and the sword.'

It is more than ironic that industries which had suffered least from fire and drought were hardest hit by the floods.

Years of drought had depleted the vineyards of South Australia. Yields were thin, the harvest in some places reduced to a sixth of normal. Hot winds from the fires further damaged the vines themselves, especially around Clare where output was 60 per cent below normal. The storms caught the wine industry, already reconciled to a 30 per cent drop in output, in the middle of harvest.

Vintners in the Barossa were accustomed to disasters; in November 1979 freak hail storms devastated the vines at the beginning of the summer. But none had experienced devastation on this scale. In a single night mud washed down by flash floods from towns higher up the valley immersed vines, buildings and roads in a gluey mire 3 feet thick. Two hundred houses were wrecked and vineyards and equipment were buried under silt.

Few buildings in the Barossa were constructed to withstand the two disasters of cyclone and flash flood. At Yalumba guttering collapsed, flooding the refrigeration plant. It burned out. The underground bottling hall at Nuriootpa, with all its equipment and 150,000 dozen bottles of wine, disappeared under the silt. Many wineries which had built underground to escape the summer heat suffered the same fate.

After the storms of March statisticians began to count the cost.

Ash Wednesday's damage had already been quantified at around $200 million. Estimates of the flood damage for the Barossa Valley alone were set at $10 million and for the state as a whole at much more. The South Australian government grimly foreshadowed a deficit of $105 million – a 75 per cent increase, made up largely of the $20 million paid out in fire assistance and an estimate of relief necessary after the floods.

In lost crops and stock, in government underwriting and assistance, above all in the damage to pasture, the drought alone had cost between 70,000 and 90,000 jobs and led to a 2 per cent fall in the gross national product. In money terms the drought cost Australia something between $400 and $600 million.

On top of this came the hidden losses, difficult to assess, but dramatised for every Australian by higher prices, business dislocation and a tentativeness in new investment. Among the first to suffer were those traditional victims: the arts. The film industry in particular, despite a glamorous image and hefty government incentives to investors, had its worst year in a decade.

State taxes and insurance premiums would obviously rise as the two sectors recouped Ash Wednesday's losses. The futures industry suffered substantially, as did the property business. Newer industries stood to lose most because of the ecological assault. Rice-growers alone claimed a loss of $40 million. Recognising the unreliability of the Australian climate they had planted on the basis of receiving only as much rain as had fallen in the worst year recorded. When the situation turned out to be far worse, farmers lost an estimated 24 per cent of their crops.

The slump was repeated across the continent. It seemed impossible that anything else, aside from a plague of locusts or an outbreak of boils, could possible assault an exhausted Australia.

But El Nino still had one last trick to play.

# TWENTY-SIX

## *The floods*

PHOTOGRAPHER BOB Finlayson lifted his camera to start taking pictures as the Royal Australian Air Force Caribou helicopter swooped towards the desert of coffee-coloured water that stretched from horizon to horizon. Isolated in the gigantic lake was a tiny pancake of land, fretted at the edges by the encroaching floods. Only a few hundreds yards around, the bare patch of earth crawled with movement as the helicopter dropped and the crew prepared to heave out the bales of hay that were its sole cargo.

Finlayson's camera caught the hapless survivors marooned on the treeless, featureless island: fifty Hereford cows and twenty-five calves, forty kangaroos and thirty hares.

It was impossible to know how long they had been there. This area of northern New South Wales had been under water for three weeks. It would be another three weeks before the waters receded sufficiently to walk the stock off this island. The equivalent of $110,000 in hay had already been dropped to stranded cattle and sheep in the area, but the eight-times-weekly trips of the Caribou would shortly cease.

Finlayson's picture appeared on the front page of the *Australian* for July 22nd. For many people in Australia's cities it was the first dramatic evidence of a fact people in Queensland and northern New South Wales had lived with since the drought-breaking rains of March. El Nino's final assault on Australia was a plague of floods.

Pouring into the vacuum created by the disappearance of the high-pressure area, moist air from the north and east first soaked, then inundated the border area where Queensland and New South Wales join. More rain fell in southern Queensland in a month than was normal in an entire year. The wettest winter in twenty years was also the wettest of the century.

Children soon got tired of playing in puddles. The glamour of driving around in the pouring rain palled. People became accustomed to swollen creeks, roads cut by streams or flooded as the rivers overran their banks. The particular depression brought on by monsoon-type rains manifested itself, accentuated by the irritants of burgeoning humidity: mouldy shoes, bawlky cars, mildew and rot.

Many people from the border areas saw 1983 as a melancholy replay of 1954 and 1955, when cyclones dumped billions of gallons of water on Queensland, bursting the rivers and turning north-western New South Wales into a lake. Like 1982, 1953 had been a year of drought and, again as in 1983, the drought's breaking had been excessive and catastrophic with more rain falling in three days of January 1954 than during the whole of the previous year.

The damage both to coastal towns and inland farming areas in 1954 had been intense, with 10,000 homes flooded and twenty-two people killed. Film of the northern New South Wales town of Maitland, with a river rushing through the main street and desperate residents clinging to rooftops, became one of the most vivid images of Australian environment in revolt. It retains its violence to this day and inspired a striking sequence for the successful Australian film *Newsfront* in 1978. Ironically 1954 was also a royal visit year; Queen Elizabeth had the chance to see at first hand the same degree of ecological ruin that was viewed by her son and daughter-in-law almost thirty years later.

Through the autumn and early winter of 1983 the water rolled south into the slopes and plains to the west of the Great Dividing Range, looking for escape to the sea.

Finding none it bloated, then burst the banks of the Darling, which flows south-west until it joins the Murray at the Victorian border.

The water's steady, inexorable progress highlights Australia's sheer size. Though the Queensland rains fell mostly in May, it was June 5 before flood waters reached Walgett in northern New South Wales and experts estimated that a further two months would pass before they reached the confluence of the Darling and Murray 1000 miles south-west. When the water finally ran into South Australia's Lake Alexandrina, they would have travelled almost 2500 miles.

After four years of drought even the mightiest of Australia's rivers were clogged with topsoil washed down by erosion, their mouths dammed by silt and mudflats. At the ocean-mouths of smaller rivers, salt water had penetrated so far inland that holiday fishermen in towns near the coast watched in horror as sea-going sharks prowled past their lines.

Since December 1981 no water had flowed out of the huge Murray River except during a freak three days in July 1982; on July 1st, 1983 the flow began again, avatar of the huge volume gathering itself thousands of miles to the north.

For decades, flood control had been a pawn in the political chessgame. Proponents of dams which would pour more water into these coastal streams (and their newly crowded and thus politically attractive towns) wrestled for support against farming interests anxious to drive tunnels through the mountains, diverting what water there was west, into the arid centre of the state that had been transformed from lush farming land to semi-desert by ruthless over-cultivation and the slow passage of Australia towards a drier climate.

By April 1983, however, nothing had been settled and the people of cities like Grafton, on the coastal side of the mountains, began sandbagging the banks of the Clarence in

expectation of inundation, while inland farmers daily track-
ed the floods' progress as they fanned out across the
drought-hardened plains, turning an area of 5000 square
miles into what Nigel Austin called in the *Australian*: 'a
tiara of archipelagoes and scattered islands'.

Farmers in Australia are as used to flood as they are to
fire; they know it as a force that, again like fire, brings in its
wake fertility and regeneration of the environment. After a
rain storm, even the bleached inland desert blooms, as
wildflower seeds, dormant for years, seize their brief
opportunity to propagate.

As graziers struggled to feed their stock, they were
planning for the spectacular year they knew must follow,
when the parched land, soaked with water, would produce
furious growth in crops like cotton, lush grazing for stock
and three years of prize-winning herds.

And after that? Nobody much cared to speculate. In
Australia nature always had something up its sleeve.
Another El Nino could not be expected until 1990, but that
of 1983 had, after all, not run exactly according to rule, so a
new drought might strike at any time. Or perhaps Australia
would be vexed by the plague of locusts jokingly predicted
by victims in the Barossa Valley; Australia had suffered
from grasshopper plagues before and would no doubt do so
again.

During the late seventies, rabbits overran large areas of
semi-arid country. Now kangaroos were once again becoming a pest and there were moves, after the fuss of Ash
Wednesday had died down, to embark on a national extermination campaign against them, with extensive federal
government support. There was always something. For
Australia fighting the environment had become in itself a
way of life.

## TWENTY-SEVEN

### *Ashes to ashes*

AT THE end of May all that remained of Ash Wednesday was, ironically, the ashes. Even the media ceased to take any serious interest in the event or its aftermath now that it was time to seek causes and lessons for the future.

Not that anyone showed an eagerness to look for lessons. It was a time for inquests and lawsuits, for excuses and accusations, the 'persecution of the innocent and the rewarding and promotion of the incompetent'. Victorian police sergeants Ian Findlay and Graham Newbegin who, it was alleged, had forcibly evacuated two men trying to defend their house against a firefront were reported to be under investigation. Their chief inspector remarked phlegmatically: 'It would appear that a member of the public who owes these men a great deal is showing his ingratitude by complaining.' Neither man could receive the awards voted to him by grateful private citizens until the investigation was over.

In wintry Cockatoo and Macedon tempers were short. For people with no insurance or for those whose claims were in dispute, there was only the long, hard wait for help. 'The adrenalin has run out and now it's just a long, hard slog,' said a community worker.

Most lived in caravans, crowded on to sites increasingly ignored by those more fortunate. According to a *Sun* report, 'Men are drinking heavily, tension is straining family relations and the problems of living through winter

181

in makeshift homes is frustrating.' Some complained that shoes issued to them by relief agencies were causing foot problems – a reminder that the low morale of fighting men can often be gauged from the incidence of such minor irritations. For the people of Cockatoo and other burned-out towns it had been a long campaign with no victory in sight.

But a victory of sorts was won on July 7th. Pressed for an acknowledgment of its role in the fires of Ash Wednesday, Victoria's State Electricity Commission announced that it expected to pay out 'something short of $100 million' on the 1019 claims filed with the commission itself and the 102 writs then before the Supreme Court in which people in the Macedon–East Trentham area charged that the fires of Ash Wednesday had resulted from the failure to maintain power lines.

While offering to pay 'reasonable claims' the SEC was insistent on accepting no liability for the fires of February. In the face of this apparent absurdity, those whose houses had been destroyed followed a line of reasoning no less baffling. 'I don't care if the SEC denies liability,' said a lawyer for one group of litigants, 'as long as it pays up.' This apparently strange indifference to the liability of the SEC is of course understandable when one recalls that the primary responsibility of a lawyer is to his clients and their interests must take priority over any pursuit of an institution.

Nobody seemed concerned that the SEC had 'paid up' before and would presumably have to pay up again in the future if courts found its antiquated methods of power supply responsible for further bushfires. The possibility of a final solution to the problems of a faltering electrical system was conveniently pushed under the mat, where it would smoulder until the next bushfire season.

Elsewhere, war with the environment once again moved into high gear.

In April kangaroo-killing became an issue once more. The American Fish and Wildlife Service reimposed a ban on the import of kangaroo skins and meat into the USA. The Australian government and industry alike rose to defend the slaughter in the interests of husbandry. The government sent to the United States two experts, the assistant director of the New South Wales National Parks and Wildlife Service, and the Primary Industry Department's director of information. Their job was to brief Australian consular officials on what the *Sydney Morning Herald* euphemistically called 'the kangaroo problem and management system'.

Most concern centred around a film called *Goodbye Joey* in which kangaroo hunters were shown torturing and mutilating the animals. Both visiting experts were scheduled to appear with representatives of the Sierra Club and other American environmental groups after an NBC screening of the film. There was some doubt about the objectivity of *Goodbye Joey* and Australian stations refused to run the film. There was, however, little doubt that not only every brutality shown in the film but others much worse had been inflicted on kangaroos and would continue to be by farmers who despised and resented these gentle and greedy animals.

The Australian government argued that none of the three varieties of kangaroos – red, eastern grey and western grey – were 'endangered', and that 'controlled harvesting' was the best means of ensuring their preservation. Ecologists were heard to murmur that there were more sensible ways of dealing with the problem than mass extermination. Did nobody recall that a similar war of attrition against Australia's flora would leave the continent in its bicentennial year of 1984 with only a third of the trees that existed when the white man came?

The success of environmental campaigns against the damming of Tasmania's Franklin River briefly gave conservationists the illusion they were winning the battle but that furore soon died. A troublesome lobby had been 'paid off' just as the Macedon litigants had been paid off. And, for most Australians, that was enough.

When everyone's claims have been settled, when the houses are rebuilt and the vineyards and paddocks are green again, it is this typically Australian emphasis on being paid and getting back to work that will guarantee another Ash Wednesday in five or seven years' time.

Research into solutions to Australia's ecological impasse will continue to be underfunded. Despite the unexpected victory of the anti-Franklin Dam campaigners expedience will continue to be the rule by which all environmental decisions are made.

It is expedient to deny that anything can be done about bushfires, to place the responsibility on householders and town councils and not to impose upon organisations like electricity commission safety regulations that in other countries are seen as simple common sense. No administrative body, from the federal government down to the smallest shire council or CFS group, wants the responsibility or the unpopularity of enforcing rules for the protection of all.

What are the lessons of Ash Wednesday?

The most important is that fires cannot be avoided. Australia is trapped in an ecological fire zone. The bush must burn in order to survive. The effect of fires can be minimised, however, the loss of life and property vastly decreased by making electrical wiring safe, forcing householders to clear the areas around their homes, cutting adequate fire trails and providing water sources for firefighting tankers.

Fire service volunteers desperately need to be more adequately protected from the possibility of being trapped in a fire storm. The emergency services' disaster plan . . .

But already one hears the conventional arguments: cost, loss of amenity, the problems of administration and control. There is no memory so short as that of a bureaucrat for the errors of his predecessor, no concern more casual than that of an Australian householder over what happened last month, let alone last year. There is always, after all, the anodyne of barbeque, bar and beach – sovereign remedies for those twinges of social responsibility or historical perspective.

Somewhere in the middle of all this the result was announced of a $10,000 competition for a bushfire-resistant house. The winning design had shutters to exclude sparks, rounded edges and gutters that would not collect debris. It was constructed of – what else? – wood.

# INDEX

188

192